Reviews and critical articles covering the entire field of normal anatomy (cytology, histology, cyto- and histo-chemistry, electron microscopy, macroscopy, experimental morphology and embryology and comparative anatomy) are published in Advances in Anatomy, Embryology and Cell Biology. Papers dealing with anthropology and clinical morphology that aim to encourage co-operation between anatomy and related disciplines will also be accepted. Papers are normally commissioned. Original papers and communications may be submitted and will be considered for publication provided they meet the requirements of a review article and thus fit into the scope of "Advances". English language is preferred, but in exceptional cases French or German papers will be accepted.

It is a fundamental condition that submitted manuscripts have not been and will not simultaneously be submitted or published elsewhere. With the acceptance of a manuscript for publication, the publisher acquires full and exclusive copyright for all languages and countries.

Twenty-five copies of each paper are supplied free of charge.

Manuscripts should be addressed to

Prof. Dr. F. **BECK,** Department of Anatomy, University of Leicester, 6 University Road, GB-Leicester LE1 7RH

Prof. W. **HILD,** Department of Anatomy, Medical Branch, The University of Texas, Galveston, Texas 77550/USA

Prof. Dr. W. **KRIZ,** Anatomisches Institut der Universität Heidelberg, Im Neuenheimer Feld 307, D-6900 Heidelberg

Prof. Dr. R. **ORTMANN,** Anatomisches Institut der Universität, Lindenburg, D-5000 Köln-Lindenthal

Prof. J.E. **PAULY,** Department of Anatomy, University of Arkansas for Medical Sciences, Little Rock, Arkansas 72205/USA

Prof. Dr. T.H. **SCHIEBLER,** Anatomisches Institut der Universität, Koellikerstraße 6, D-8700 Würzburg

E.B. Krammer M.F. Lischka T.P. Egger
M. Riedl H. Gruber

The Motoneuronal Organization of the Spinal Accessory Nuclear Complex

With 11 Figures

Springer-Verlag Berlin Heidelberg GmbH

Eva B. Krammer
Martin F. Lischka
Thomas P. Egger
Maria Riedl
Helmut Gruber

Institute of Anatomy, University of Vienna
Währingerstr. 13, A-1090 Vienna

ISBN 978-3-540-17459-2   ISBN 978-3-662-10362-3 (eBook)
DOI 10.1007/978-3-662-10362-3

Library of Congress Cataloging-in-Publication Data
The Motoneuronal organization of the spinal accessory nuclear complex. (Advances
in anatomy, embryology, and cell biology; v. 103) Bibliography: p.   Includes index.
1. Accessory nerve—Anatomy. 2. Motor neurons. I. Krammer, E.B. (Eva B.),
1944–   . II. Series. [DNLM: 1. Accessory Nerve—anatomy & histology. 2. Accessory Nerve—Physiology.
W1 AD433K v.103/WL 330 M919]   QL801.E67   vol. 103   574.4 s   86-33868
[QM471]   [599′.04′8]

© Springer-Verlag Berlin Heidelberg 1987

Originally published by Springer-Verlag Berlin Heidelberg New York in 1987.

The use of general descriptive names, trade names, trade marks, etc. in this publication, even if the former are not especially identified, is not to be taken as a sign that
such names, as understood by the Trade Marks and Merchandise Marks Act, may
accordingly be used freely by anyone.

Product Liability: The publisher can give no guarantee for information about drug
dosage and application thereof contained in this book. In every individual case the
respective user must check its accuracy by consulting other pharmaceutical literature.

2121/3140-543210

Acknowledgments

The senior author wishes to dedicate this study to Prof. Dr. h.c. mult. Paul A. Weiss and to Prof. Dr. Benno Schlesinger, who have taught her many things.

The authors greatly appreciate the expert technical assistance given by Ms. Rikki Schramm. Several people who have helped with the photography, including Mr. W. Bruckner, Mr. Yeti, and Ms. Jeanette Beinl, are gratefully acknowledged. It is a pleasure to extend thanks to Mr. Simon North for his conscientious scrutiny of the English of the manuscript. The authors are very appreciative of the financial support given by Österreichische Forschungsgemeinschaft, Förderungsprogramm "International Communication."

Contents

Abbreviations

$C_{1(2,3\ldots7)}$ 1st (2nd, 3rd ... 7th) cervical segment or ventral ramus of cervical nerve

C1 (2,3 ... 7)   caudal half of the 1st (2nd, 3rd ... 7th) cervical segment

cbCP   communicating branches of the cervical plexus

CLM   cleidomastoid muscle

CLMn   cleidomastoid nerve

CLO   cleido-occipital muscle

CLOn   cleido-occipital nerve

CP (3 + 4)   cervical plexus (3rd and 4th ventral rami)

R1 (2,3 ... 7)   rostral half of the 1st (2nd, 3rd ... 7th) cervical segment

spXIn   spinal accessory (XI) nerve

STM   sternomastoid muscle

STMn   sternomastoid nerve

TRAP   trapezius muscle

TRAPn   trapezius nerve

TRAPr   trapezius ramus of the spinal XI nerve

1 Introduction

Though more than 300 years have elapsed since the first description of the peculiar course of the spinal accessory (XI) nerve by Willis (1664), the crucial problems concerning what is known as accessory field of musculature and its innervation are still unsolved and a matter of controversy. Like the bulbar XI, the spinal XI nerve is commonly regarded as originally a branch of the vagus and, therefore, as a cranial nerve (Fürbringer 1897; Gegenbaur 1898; Lubosch 1899). However, whether this nerve is of special visceral or somatic derivation is still debated. The conventional distinction between these functionally separate categories of cranial nerves is based largely on two criteria, namely, the position of the cranial nerve nucleus and the embryological derivation of the muscles innervated by this nerve. Unfortunately, little is known about the development of this accessory field of musculature, and the evidence concerning the position of the spinal XI nucleus is contradictory. In fact, although the spinal XI nerve is usually regarded as a purely efferent nerve belonging to the special visceral efferent group of cranial nerves and innervating muscles derived from the branchial mesoderm, each of these properties has been questioned. Consequently, the classification of the nerve is still unsettled.

Evidence in support of a special visceral origin of the spinal XI nerve is found in the phylogenetic history of the spinal XI nucleus. Thus, the motoneurons of the spinal XI nerve are generally considered to have migrated from the motor nuclear complex of the vagus in the medulla oblongata into the cervical portion of the spinal cord (Fürbringer 1897; Gegenbaur 1898; Lubosch 1899; Black 1917a, b, 1920, 1922; Ariëns Kappers et al. 1936; for review see also Straus and Howell 1936). A minority of authors, however, regard the spinal XI nerve to be of spinal origin or at least as a nerve of the transitional territory of medulla and spinal cord, the reason being that it arises from large motoneurons lying in the caudal prolongation of the somatic hypoglossal nucleus. Hence, the spinal XI nerve is classified by some (Beccari 1913, 1914; Addens 1933) as a somatic efferent nerve.

As there is also no agreement concerning the origin of the spinal XI musculature, the derivation of these muscles can hardly be adduced to classify the spinal XI nerve. Although most authors (Allis 1897; Fürbringer 1897; Gegenbaur 1898; Lewis 1910; Edgeworth 1911, 1926; Luther and Lubosch 1938) hold that these muscles are derived from the mesoderm of the branchial arches, some (Favaro 1903; Addens 1933) consider them to develop at least partially from occipital and cervical myotomes. From the literature (e.g., Addens 1933), we would conclude that these muscles are probably of mixed somitic and branchial origin. Therefore one would anticipate the motor innervation of these

1

muscles to be mixed, that is, both somatic and visceral, and to have certain characteristics of both categories. In higher vertebrates, i.e., sauropsides and mammals, the spinal XI musculature is indeed innervated by cervical nerves as well as by the spinal XI nerve (Addens 1933; Straus and Howell 1936). The double innervation of these muscles has never ceased to puzzle anatomists and has given rise to many controversies. On the assumption that the cervical branches contribute motor fibers to the spinal XI muscles, those parts of the muscles supplied by cervical nerves are considered to be of myotomic origin, while those parts innervated by the spinal XI nerve are thought to be of branchiomeric origin (see e.g. Black 1917a, b; Addens 1933; Straus and Howell 1936). Accordingly, it is supposed that the cervical nerve innervation has been added to these muscles, as trunk myotomes have contributed to their formation. However, there is strong evidence that the spinal XI muscles receive their entire motor innervation from the spinal XI nerve, whereas the communicating branches of the cervical plexus supply only proprioceptive fibers to them (Straus and Howell 1936; Corbin and Harrison 1938; Yee et al. 1939). If this assumption proves to be true, the spinal XI nerve has to be regarded as a purely efferent nerve receiving muscle sense fibers from cervical ventral rami via peripheral anastomoses (Hinsey and Corbin 1934; Straus and Howell 1936; Corbin and Harrison 1938; Yee et al. 1939).

In order to leave no doubt about the complexity and obscurity of the problem, it seems appropriate to point out that in addition, a sensory component appears to be present in the spinal XI nerve in many mammalian embryos, including human embryos (Windle 1931a, b; Pearson 1938). These afferent fibers of the spinal XI nerve have been shown to arise in small ganglia and from scattered groups of ganglionic cells, having been found along the intra- and extracranial course of the spinal XI nerve (Weigner 1901; Streeter 1905; Fahmy 1927; Windle 1931b; Pearson 1938; Kimmel 1940; Waibel 1954; Pearson et al. 1964). Moreover, additional sensory fibers are believed to arise in the dorsal root ganglia of the upper cervical nerves, since connections between the latter and the spinal XI trunk have been demonstrated (Kazzander 1891; Lubosch 1899; Weigner 1901; Streeter 1905; Windle 1931b; Pearson 1938; Kimmel 1940; Pearson et al. 1964). According to the presence of sensory fibers and ganglia in the spinal XI trunk, the spinal XI nerve would have been classified as a mixed nerve (Windle 1931b). Again, in this case, the question remains: why is there an additional supply of the spinal XI muscles by sensory fibers of the cervical plexus? The fact that the spinal XI nerve tends to lose its sensory fibers in the course of development (Streeter 1905; Straus and Howell 1936) could explain their replacement by those of cervical branches. On the other hand, there is no doubt that the occurrence of sensory fibers in embryos, arising from ganglion cells along the spinal XI trunk and joining the solitary tract (Darkschewitsch 1885; Windle 1931a, b; Pearson 1938; Pearson et al. 1964), suggests the presence of a visceral afferent component in this nerve, thus indicating its visceral derivation. This point of view is further supported by the fact that the emergent course of the spinal XI roots closely resembles that of visceral motor fibers arising in the IXth and Xth cranial nuclei (see e.g., Lubosch 1899; Kimmel 1940; Crosby et al. 1962).

Thus, despite the large number of findings and the conclusions based on them, two fundamental questions concerning the spinal XI nerve still require

2

solution, namely, its classification and the significance of the communicating branches of the cervical plexus.

Although the position of a cranial nerve nucleus is one of the main criteria for classifying cranial nerves, there is not even unanimity with regard to the exact localization of the spinal XI nucleus in the cervical cord. Some authors believe that it lies medially in the ventral horn (Henle 1871; Hogg 1928), but most believe they have identified it somewhere in the lateral part of the ventral horn of the cervical cord (Darkschewitsch 1885; Koelliker 1896; Lubosch 1899; Gehuchten 1900; Black 1917a, b, 1920, 1922; Straus and Howell 1936; Pearson 1938; Kimmel 1940; Rexed 1952, 1954; Pearson et al. 1964). A small group of investigators assume a medial in addition to a lateral origin of the spinal XI nerve (Hepburn and Waterston 1904; Angulo y González 1927; Rapoport 1978), while others state that the spinal XI nucleus changes from a medial to a lateral position throughout its rostrocaudal extents (Kaiser 1891; Bruce 1901; Yoda 1940; Romanes 1941; Matsushita 1970; Silver and Wolstencroft 1971; Holomáňová et al. 1972, 1973; Gura and Limanskii 1977). In addition, a variety of longitudinal extensions of the spinal XI nucleus are reported in different species and in the same species. As most of these results are based merely on cytoarchitectonic studies, the discrepancies in the location and position of the spinal XI nucleus are most probably due to difficulties in identifying this nucleus by solely descriptive methods. Little information on the location of the spinal XI nucleus has been provided by experimental studies (Straus and Howell 1936; Holomáňová et al. 1972, 1973; Gura and Limanskii 1977; Rapoport 1978), while even less is known about the somatotopic organization of the motoneurons within the spinal XI nucleus. Furthermore, there have been only a few attempts utilizing the technique of retrograde transport of horseradish peroxidase (HRP) to localize motoneurons innervating individual spinal XI muscles within the cervical cord (Rapoport 1978; Gottschall et al. 1980b; Robards et al. 1980; Karim and Hoo Nah 1981). However, some anatomical and methodological aspects were not taken into consideration in the only experimental study on the somatotopic organization of the spinal XI nucleus in the cat (Rapoport 1978). Thus, injection of HRP directly into a muscle, a technique employed by Rapoport (1978), was found to produce spurious labeling of motoneurons resulting from diffusion of HRP to adjacent muscular branches (Richmond et al. 1978). On the other hand, it may be that the HRP solution does not spread sufficiently to all nerve fibers innervating a particular muscle.

Further uncertainties in these experimental results arise from the fact that deductions from human myology, though a convenient simplification in somatotopic studies, are not generally applicable to the anatomy and function of muscles in quadrupedal mammals. For example, the clavicle is absent in the cat and in all animals in which the forelimbs are used primarily or entirely for progression (e.g., Ellenberger and Baum 1932; Sandstrom and Saltzman 1944). In consequence, those parts of the spinal XI muscles usually attached to the clavicle are modified in being inserted into the humerus or antebrachium of these quadrupedal mammals (see e.g., Owen 1868; Ellenberger and Baum 1932). Therefore, in the cat, these muscles are primarily concerned with quadrupedal locomotion and cannot be regarded as homologous to the corresponding muscles in man. Furthermore, the spinal XI musculature of most mammals comprises not merely two muscles, as assumed by some investigators (Rapoport

3

1978; Karim and Hoo Nah 1981), but at least four individual ones (Streissler 1900; Schück 1913; Luther and Lubosch 1938; Jouffroy 1971). Apparently, the sternocleidomastoid and the trapezius have been regarded as individual muscles, although either of them consists of two units. Accordingly, the moto-neurons of two separate muscles have been labeled in common.

In view of the uncertainty of these experimental results, we investigated the location and organization of the spinal XI nucleus in a different species. The rat was selected because of the peculiar anatomy of its shoulder girdle. Although it is a horizontally postured quadrupedal animal (DuBrul 1950), it has a clavicle like those species which use the forelimbs for prehension, e.g., many rodents, the primates, and man (Ellenberger and Baum 1932; Sandstrom and Saltzman 1944). The spinal XI musculature of the rat, consisting of at least four individual muscles (Streissler 1900; Jouffroy 1971), is well developed according to the functional adaptation of the shoulder girdle to both traction and locomotion (Sandstrom and Saltzman 1944).

Since knowledge of the special myology is an essential prerequisite to an understanding of the significance of the somatotopic organization of a motor nucleus, we studied the anatomy of the spinal XI muscles and of their innerva-tion before starting our experiments. Furthermore, unlike Rapoport (1978), we used the cut-nerve exposure to HRP in order to avoid the problem of HRP spread to adjacent nerves and to provide an equalized exposure of all axons present within a particular nerve (Richmond et al. 1978; Krammer et al. 1979).

The experiments described here were aimed at elucidating some unsolved problems concerning the spinal XI nerve, such as its classification, the signifi-cance of the communicating branches of the cervical plexus, and the functional organization of the spinal XI nucleus. In an attempt to clarify the classification of the spinal XI nerve, we examined by means of HRP labeling methods the geometric position of the spinal XI nucleus within the cervical cord, the intra-medullary course of the spinal XI fibers, and the significance of the communicat-ing branches of the cervical plexus. To substantiate the labeling results, addition-al electrophysiologic experiments were performed on the communicating branches. In particular, we were interested in the somatotopic organization of the spinal XI nucleus. The localization of the neuronal groups innervating the individual spinal XI muscles within the spinal XI nucleus was therefore determined by using the cut-nerve exposure to HRP. On the basis of this experi-mental analysis of the spinal XI nuclear complex, we propose a new concept concerning the significance of the motoneuronal organization in the cervical cord.

4

2   Materials and Methods

Male Sprague-Dawley rats weighing 200–250 g were used in this study. The animals were anesthetized with Nembutal (40 mg/kg i.p.) and given a single injection of atropine sulfate (*Heilmittelwerke*) (15 µg i.p.) to prevent aspiration of respiratory secretions.

A total of 90 animals were subjected to experiments using the technique of retrograde transport of HRP. A selected nerve – the spinal XI nerve, one of its muscular branches, or one of the communicating branches of the cervical plexus – was isolated and cut bilaterally. In the data to be presented, however, only those experiments which resulted in reliable neuronal labeling (a total of 116 cases) have been included. Crystals of HRP grade I (Boehringer-Mannheim) were applied to the proximal nerve stumps until the exposed nerve fibers were observed to turn a dark brown color. Great care was taken not to traumatize the cut nerve proximally to its exposure and to avoid HRP spread to nerve fibers of adjacent muscles.

HRP was applied (Fig. 1) to the cut ends of (1) the spinal XI nerve proximal to its ramification (34 cases), (2) the trapezius (TRAP) (15 cases), (3) the cleido-occipital (CLO) (11 cases), (4) the cleidomastoid (CLM) (10 cases), and (5) the sternomastoid (STM) (13 cases) nerves just proximal to their entrance into the corresponding muscles. HRP was further applied to the cut ends of (6) the TRAP ramus (9 cases) and (7) the common stem of the STM and CLM rami of the spinal XI nerve (13 cases), proximal to their communication with branches of the cervical plexus. Finally, the cut ends of the communicating branches of the cervical plexus, i.e., (8) the communicating ramus to the TRAP nerve (8 cases) and (9) the communicating rami to the STM and CLM nerves (3 cases), were exposed to HRP, proximal to their union with the corresponding branches of the spinal XI nerve.

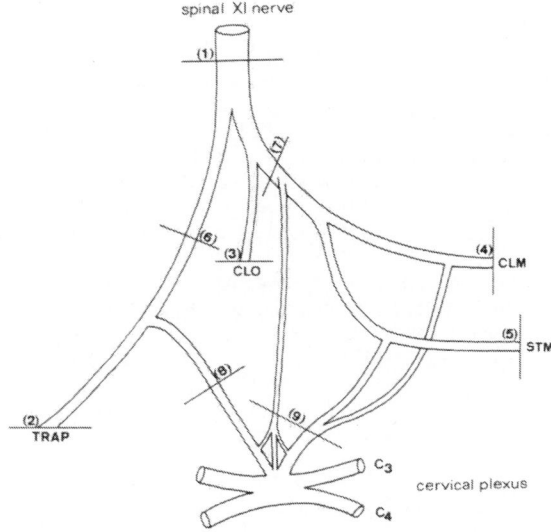

Fig. 1. Plan of the nerves supplying the spinal accessory (XI) muscles, with indication of the sites of HRP application (see Chap. 2 and Fig. 2 for further details)

After a 48–60 h survival period, the animals were reanesthetized with Nembutal (40 mg/kg i.p.), heparinized, and perfused transcardially with physiological saline followed by 1.5%–2.5% glutaraldehyde in 0.1 M cacodylate buffer (pH 7.2).

The medulla oblongata and the entire cervical spinal cord, including the dorsal root ganglia, were removed in toto and postfixed in the same fixative for 2–4 h at 4° C. After fixation the specimens were washed overnight in 0.1 M cacodylate buffer to which sucrose had been added in increasing concentrations up to 30%.

In all the experiments, the individual lengths of several cervical segments were measured prior to sectioning. The total length of a cervical segment was defined as the actual length of its dorsal root attachment plus the half-lengths of the interroot distances of both adjacent segments. Since demarcation of the 1st cervical segment is complicated either by variable reduction of the 1st dorsal roots and ganglia or by occasional persistence of an occipital (precervical) ganglion of Froriep (1882), the lowermost decussating fibers of the pyramids were taken as the rostral boundary of this segment.

After removal of the spinal ganglia, 40 µm serial sections of both the ganglia and the cord were cut in the transverse plane on a freezing microtome. The calculated number of 40 µm serial sections corresponding to the half-length of a particular cervical segment were collected and jointly processed. Tetramethyl benzidine was used as a chromogen for the histochemical detection of HRP activity (Mesulam 1976, 1978). The sections were then mounted onto chrome-alum-subbed glass slides, each of which held the serial sections of either the rostral or the caudal half of one particular cervical segment. The sections were left to air-dry overnight and counterstained with a 1% neutral red solution.

In sections selected for optimal correspondence with standard sections at rostral and caudal levels of the cervical segments from C_1 to C_7, several labeled neurons of each half-segment were recorded by camera lucida reconstruction using a Zeiss drawing tube. Accordingly, results obtained from approximately twenty-five 40 µm sections of each half-segment were entered in drawings of 14 representative transverse sections through the cervical cord.

In some animals the cervical cord, including the lower medulla, was cut into serial frontal (3 animals) and sagittal (2 animals) sections of 40 µm thickness.

An additional series of animals was used to study the anatomy of the spinal XI muscles and of the nerves supplying them. Under Nembutal anesthesia, the animals were perfused with 6.0% paraformaldehyde in 0.1 M cacodylate buffer (pH 7.2). The spinal XI muscles, the spinal XI nerve, and the cervical plexus were removed in toto. The specimens were incubated in the histochemical medium for the demonstration of acetylcholinesterase activity (Koelle and Friedenwald 1949) and stained lightly with Sudan black.

Electrophysiologic experiments were performed in rats weighing 300–400 g, which were anesthetized with Nembutal (30 mg/kg i.p.). The spinal XI nerve and those ventral rami of the cervical nerves giving off the communicating branches were dissected centrally. The decentralized nerves were mounted on bipolar platinum electrodes for stimulation and covered by a pool of paraffin oil. Stimuli, consisting of rectangular pulses of 0.7 ms duration, ranged from 1.0 to 5.0 V. Action potentials were recorded by bipolar wire electrodes positioned on the TRAP nerve. The amplified potentials were displayed on the screen of a Tektronix 565 oscilloscope and registered on 35 mm film.

In a second group of anesthetized rats, the spinal XI nerve was cut and a piece of approximately 1 cm in length excised. After a survival period of 1 week, the action potentials of the TRAP nerve were registered upon stimulation of the cervical branch communicating with it.

In several rats, the contractions of the TRAP were recorded by means of a strain gauge attached to the acromion. The vertebral origin of the superior TRAP was rigidly fixed with metal bars. Mechanography was displayed on the screen of the oscilloscope and registered on 35 mm film.

3 Results

3.1 Anatomy of the Spinal Accessory (XI) Muscles

Although the spinal XI muscular group is commonly subdivided into only two muscles – the sternocleidomastoid and the trapezius – it actually comprises at least four distinct muscles in most mammals (Streissler 1900; Schück 1913; Luther and Lubosch 1938; Jouffroy 1971). Moreover, there is a relative lack of precision concerning the anatomy and terminology of this muscular group in the rat (cf. Streissler 1900; Greene 1968; Hömig 1970; Hebel and Stromberg 1976). Therefore, a description of the muscles supplied by the spinal XI nerve is presented here prior to detailing the experimental results.

In the rat, the spinal XI nerve is found to supply the sternomastoid (STM), cleidomastoid (CLM), and cleido-occipital (CLO), as well as the superior and inferior trapezius (TRAP) (Fig. 2).

The STM arises fleshy with a superficial portion from the sternal (manubrial) crest (Klima 1973) and from the ventral surface of the rostral border of the manubrium. A deeper division arises mainly from a tendinous band which arches from the sternal end of the clavicle to the dorsal surface of the rostral margin of the manubrium. Intimately blended, both parts ascend obliquely upward across the lateral aspect of the neck, forming a thick rounded belly. A strong but narrow tendon is inserted rostral to the stylomastoid foramen into the facial (parotid) crest which, in the rat, appears to be elongated by fusion with the tympanohyal (Kampen 1905; Gaupp 1915; Frick and Heckmann 1955; Youssef 1966; Starck 1967). The STM tendon gives off a narrow slip which passes from the facial crest to the caudal end of the frontal crest, lying in a deep groove caudal to the supratympanic crest. This tendinous expansion gives attachment to the CLO, a thin narrow strap muscle which arises from the ventral surface and caudal margin of the middle third of the clavicle. It ascends parallel with the STM muscle toward its cephalic insertion.

The CLM muscle arises from the rostral border of the middle third of the clavicle. Its origin is slightly shifted toward the sternal extremity of the clavicle with reference to that of the CLO muscle. Therefore, the caudal portion of the CLM lies deep to, and is overlapped for the most part by, the CLO muscle. As it ascends almost vertically, the CLM passes deep to the STM and is under cover of the latter near its cephalic end. It is inserted at the ventral margin of the mastoid bone into a small tubercle which lies caudal to the stylomastoid foramen and probably is homologous to the mastoid apophysis (Kampen 1905; Starck 1967).

The TRAP of the rat comprises two distinct muscles, its superior and inferior parts not being fused. In addition, an occipital attachment of the superior TRAP

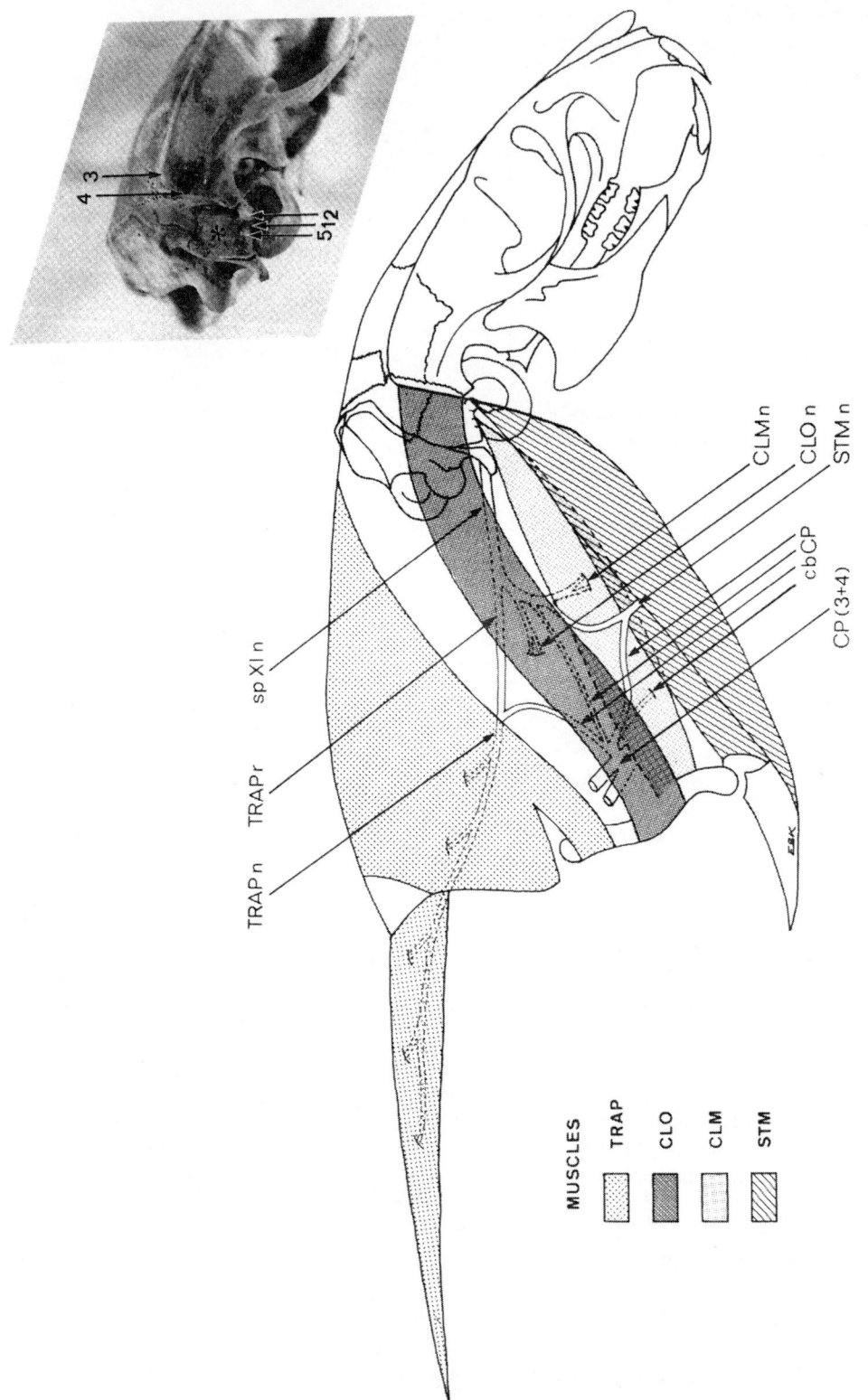

MUSCLES

TRAP
CLO
CLM
STM

CLMn
CLOn
STMn
cbCP
CP(3+4)
spXln
TRAPr
TRAPn

4 3
512

8

is absent in this species, as in many mammals (Streissler 1900). The superior TRAP arises almost exclusively from the superficial part of the ligamentum nuchae, which extends from the external protuberance of the occipital bone to the spine of the 2nd thoracic vertebra. Only the caudalmost fibers of the superior TRAP arise from the spine of this vertebra prominens. Whereas most quadrupedal mammals possess neither clavicle nor a clavicular attachment of the superior TRAP, in the rat the most rostral fibers of the TRAP are attached to the ventral surface and caudal margin of the acromial extremity of the clavicle. This insertion is covered by the omotransversarius muscle, which emerges between the CLO muscle and the clavicular part of the TRAP to pass to its acromial attachment. The middle and caudal fibers of the superior TRAP are inserted into the acromion and the crest of the spine of the scapula, respectively.

The inferior TRAP arises from the spines of the 6th to the 12th thoracic vertebrae and from the posterior layer of the thoracolumbar fascia. Its fibers converge to be attached to a small area at the medial end of the spine of the scapula.

3.2 Anatomy of the Nerves Supplying the Spinal Accessory (XI) Muscles

Under cover of the CLO muscle, the spinal XI nerve divides into a ventral and a dorsal portion (Fig. 2). The former gives off one or two filaments to the deep surface of the CLO, before it breaks up into the CLM and STM branches. Whereas the CLM ramus runs behind the CLO and CLM muscles to pierce the deep surface of the latter, the STM branch loops around the dorsal border of the CLM muscle. Passing superficial to the CLM but deep to the CLO muscles, it reaches the dorsal border of the STM muscle at the boundary of the latter's two intimately blended portions.

Emerging from behind the CLO muscle, the dorsal division of the spinal XI nerve lies comparatively superficial. Before it disappears again under the ventral border of the clavicular portion of the TRAP, it communicates with a branch from the cervical plexus to form the TRAP nerve. This TRAP nerve supplies a variable number of filaments to several portions of the TRAP muscle.

The cervical branch communicating with the TRAP ramus (of the spinal XI nerve) arises from a union of the ventral rami of the 3rd and 4th cervical nerves. Two additional communicating branches leave the same union of cervical nerves. One of them ascends to join the ventral division of the spinal XI nerve near its ramification. The second communicating branch divides at the dorsal margin of the CLM muscle into two filaments joining the CLM and STM

◁ **Fig. 2.** Semischematic drawing illustrating the spinal accessory (XI) muscles and their innervation in the rat. For identification of various structures, compare the schematic outline of the cranium with the photograph of a lateral aspect of the rat skull (inset). The tendon of the STM is inserted rostral to the stylomastoid foramen (*1*) into the facial (parotid) crest (*2*). A narrow slip of the STM tendon passes from the facial crest to the caudal end of the frontal crest (*3*), and lies in a deep groove caudal to the supratympanic crest (*4*). This tendinous expansion gives attachment to the CLO. The CLM is attached to a small tubercle (*5*), which lies caudal to the stylomastoid foramen at the ventral margin of the mastoid bone (*asterisk*). (See Sects. 3.1 and 3.2 for a detailed description)

9

rami of the spinal XI nerve, respectively. Although the way in which the communicating branches join the spinal XI rami is somewhat variable, the foregoing pattern is, however, the most constant arrangement.

3.3 The Spinal Accessory (XI) Nucleus

Following HRP exposure of the spinal XI nerve proximal to its ramification (at site (1) in Fig. 1), motoneurons of the spinal XI nucleus are found labeled throughout the upper seven segments of the cervical cord. The labeled nuclear complex of the spinal XI nerve is topographically divided into a medial and a lateral portion, both of them extending in a columnar form parallel to the long axis of the cervical cord (Figs. 3, 4, 5a, b).

The medial subnucleus of the spinal XI nuclear complex starts as a small group of cells in the transitional zone between spinal cord and medulla oblongata at the lower parts of the motor decussation (see Fig. 3a and R1 in Fig. 4). Occasionally, scattered neurons can be found either among crossing bundles of the motor decussation or descending sulcomarginal bundles. At most rostral levels of the 1st cervical segment (C_1), where only an upper prolongation of the medial motoneuronal group represents the ventral horn, the labeled neurons of the medial XI subnucleus are sparsely distributed along its dorsomedial border. A gradually increasing and distinct medial subnucleus with tight clustering of labeled neurons is present in the caudal half of the 1st and the rostral half of the 2nd cervical segments (see Fig. 3b, c and C1 and R2 in Fig. 4). At these levels, the medial subnucleus occupies a mediodorsal position within the ventral horn. However, its neurons appear to withdraw progressively from the medial margin of the ventral horn and to shift more toward the interior of the gray matter, proceeding from rostral to caudal. In the caudal half of the 2nd cervical segment, the neuronal density of the medial subnucleus decreases again, although the area occupied by the spinal XI motoneurons remains constant (see Fig. 3d and C2 in Fig. 4). In C_3, particularly in its rostral half, scattered labeled neurons appear to change over a central to a more lateral position within the ventral horn (see Fig. 3e, f and R3 and C3 in Fig. 4). No labeled neurons of the medial subnucleus are found below the caudal levels of C_3.

The lateral subnucleus of the spinal XI nuclear complex starts at rostral levels of the 2nd cervical segment, with few neurons lying at the lateral margin of the ventral horn (see R2 in Fig. 4). This subnucleus is located dorsal with respect to the ventromedial motoneuronal group, but slightly ventral as compared with the medial subnucleus of the spinal XI nuclear complex (see especially Fig. 3e and R3 in Fig. 4). The diameter of the subnucleus and its neuronal density increase in the caudal half of the 2nd cervical segment. At rostral levels of C_3, the neuronal density of this subnucleus increases substantially and reaches a maximum which is continued caudally into rostral levels of the 5th cervical segment (see Fig. 3e–j and R3–R5 in Fig. 4).

Fig. 3a–o. See pp. 12–14

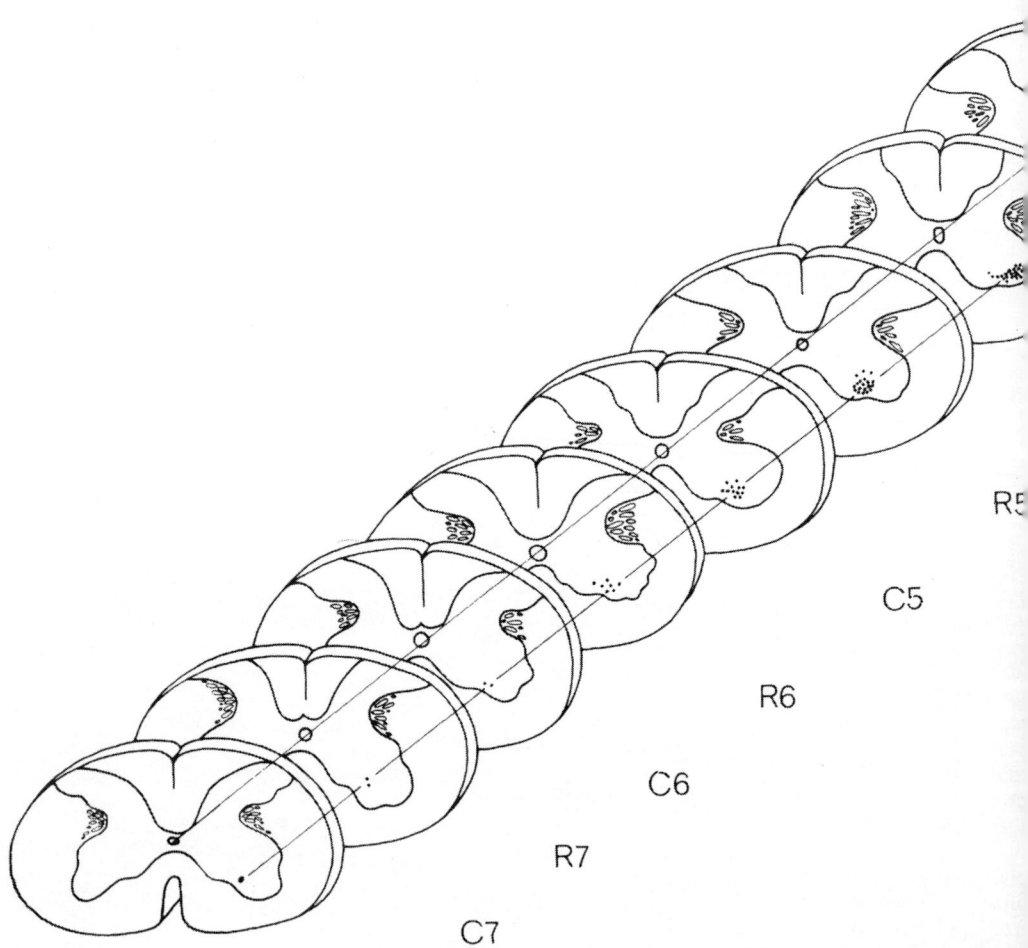

R5

C5

R6

C6

R7

C7

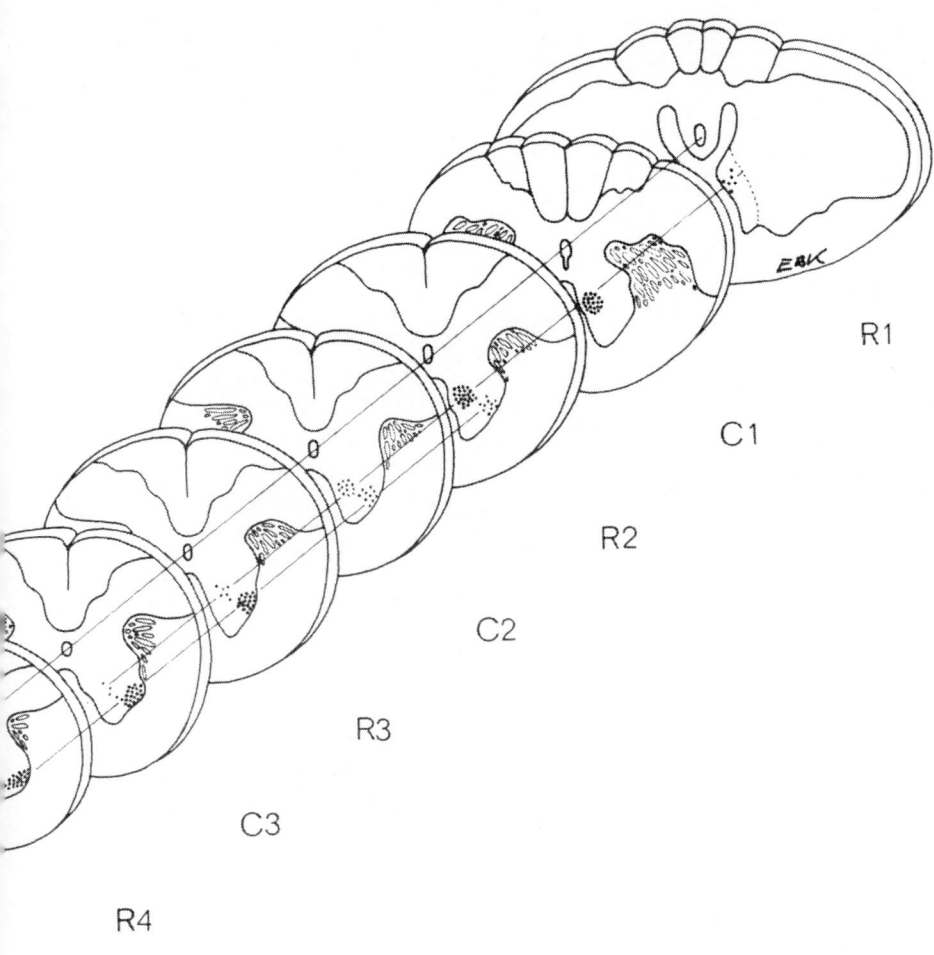

R1

C1

R2

C2

R3

C3

R4

C4

Fig. 4. Location and distribution of labeled motoneurons of the spinal accessory (XI) nuclear complex following HRP exposure of the spinal XI nerve proximal to its ramification. Camera lucida drawings of transverse sections through the cervical cord, selected for optimal correspondence with standard sections at rostral (*R*) and caudal (*C*) levels of cervical segments $C_1(1)$–$C_7(7)$. Results obtained from approximately twenty-five 40 μm sections of each half-segment are entered in drawings of 14 representative transverse sections. The serial reconstructions are arranged in a perspective view from caudal (*bottom left*) to rostral to illustrate that the long axes of both spinal XI subnuclei lie parallel to the longitudinal axis of the central canal and to each other throughout their rostrocaudal extents. The medial subnucleus of the spinal XI nuclear complex extends from R1 to C3, and its lateral fellow is present from R2 to C7. Although the subnuclei occupy different positions within the ventral horn, their facing borders appear to fuse at the caudal extremity of the medial subnucleus at C3. Note that both subnuclei are spindle-shaped, with loosely arranged neurons at their tapering extremities and tightly packed motoneurons at those intermediate portions with larger diameters. Note in particular that the progressive shortening of the ventral tip of the ventral horn below R3 causes a relative displacement of the lateral subnucleus into a ventral direction. Below C4 the concurrent lateral extension of the ventral horn, accommodating brachial plexus neurons, appears to push the lateral subnucleus from its ventrolateral position medialward (compare with Fig. 3 and see Sect. 3.3 for further details)

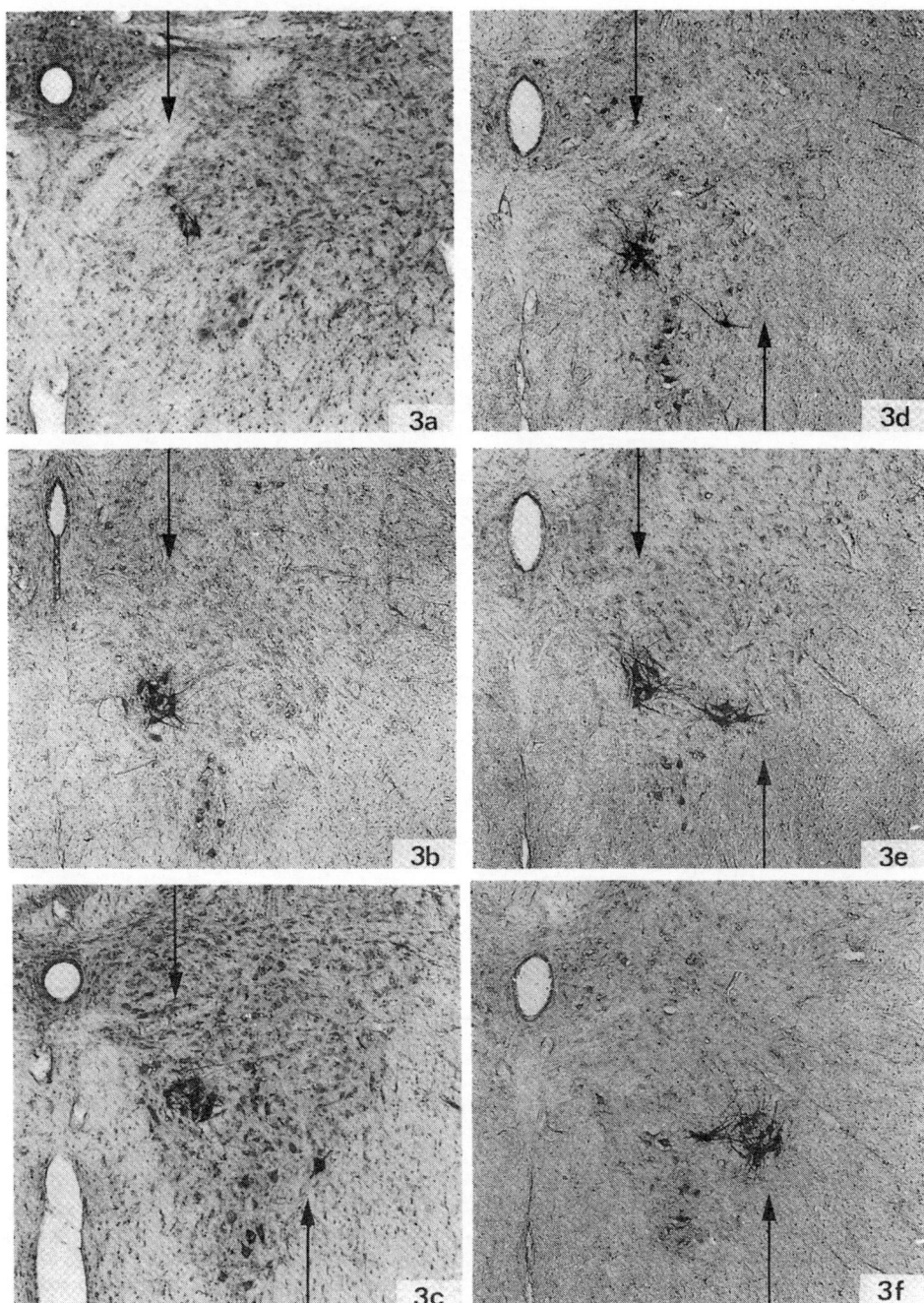

Fig. 3a–o. Rostral (**a**) to caudal (**o**) series of transverse sections through the cervical cord to show the location of the spinal accessory (XI) nuclear complex labeled with HRP following exposure of the spinal XI nerve proximal to its ramification. **a** Medullospinal transition at caudal levels of the motor decussation. **b** Caudal level of C_1. **c** Rostral and **d** caudal levels of C_2. **e** Rostral and **f** caudal levels of C_3. **g–ö** see pp. 13/14

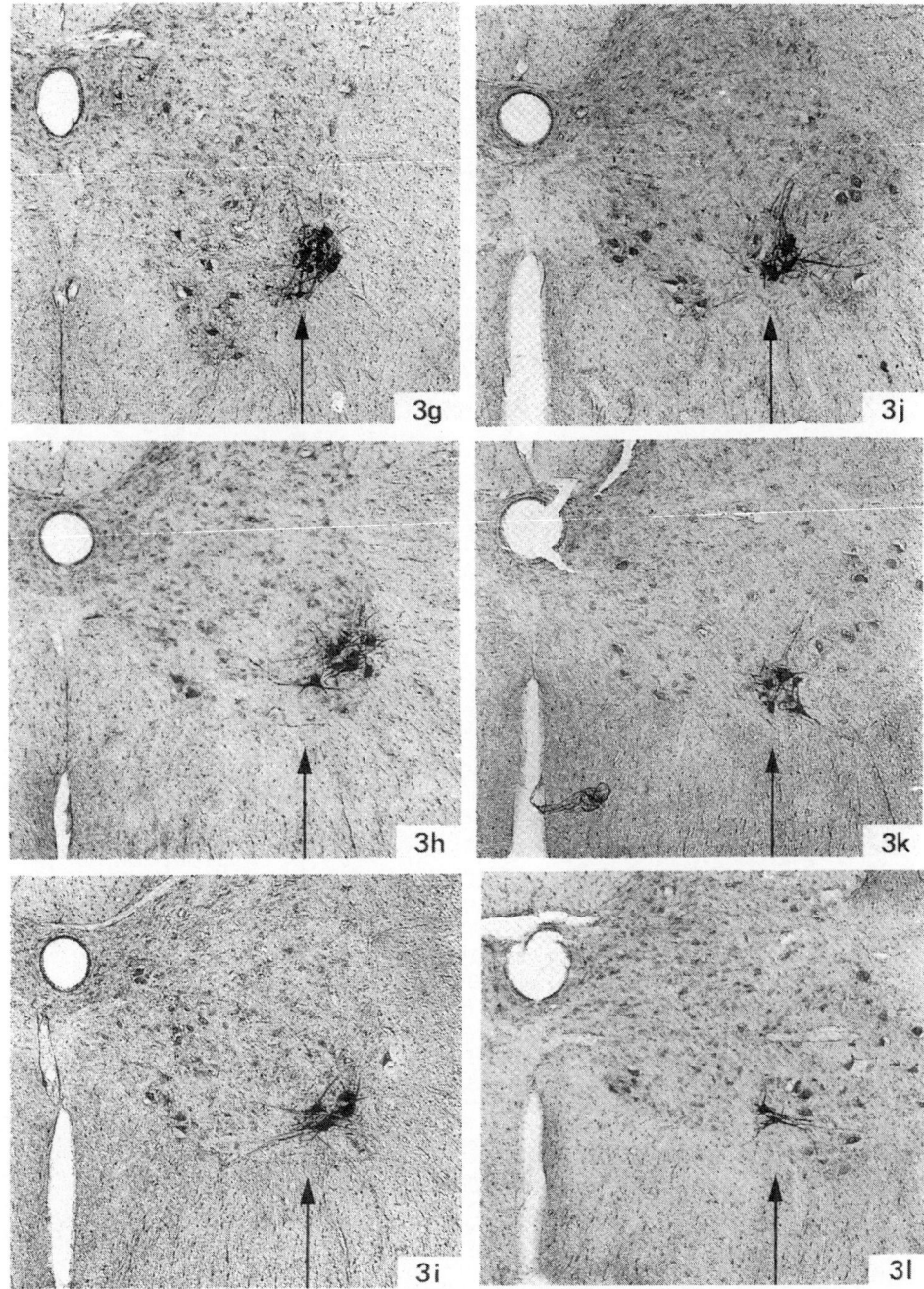

Fig. 3g–l. g Junction of C_3 and C_4. **h** Rostral and **i** caudal levels of C_4. **j** Rostral and **k** caudal levels of C_5. **l** Rostral level of C_6. **m–o** see p. 14.

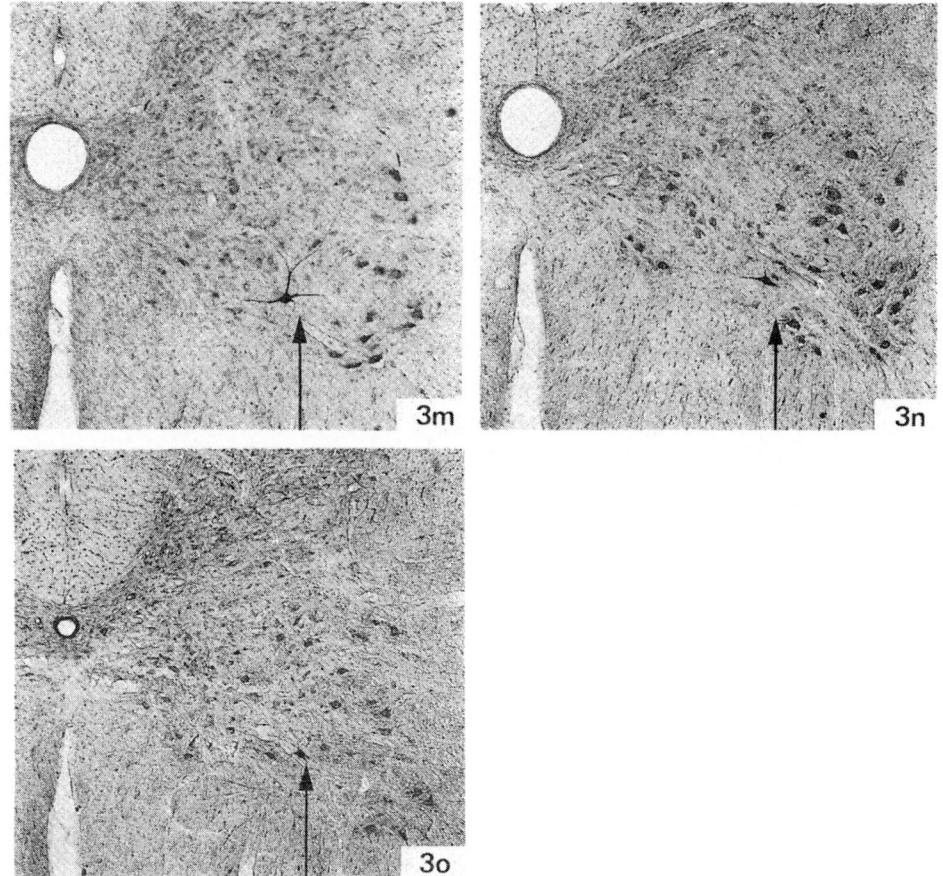

Fig. 3m–o. m Caudal level of C_6. **n** Rostral and **o** caudal levels of C_7. The spinal XI nuclear complex extends from the transitional zone between the medulla oblongata and the spinal cord (**a**) to caudal levels of C_7 (**o**). It comprises a medial (↘) and a lateral (↗) subnucleus which overlap in a longitudinal direction and fuse at the caudal extremity of the medial subnucleus (**e** and **f**). Note in particular the constant location of both subnuclei with reference to the central canal and the ventral median fissure, although changes in the configuration of the ventral horn convey the impression of a positional shift of the spinal XI subnuclei throughout their rostrocaudal extents (see Sect. 3.3 for a more detailed description) × 50

Whereas from caudal C_2 to rostral C_3 levels the lateral subnucleus occupies a position within the ventral horn similar to that already observed at rostral C_2, an apparent shifting of the subnucleus takes place at caudal levels of the 3rd cervical segment. Thus, apparently coupled with a progressive shortening of the ventral tip of the ventral horn at caudal C_3 levels, there is a displacement of the lateral subnucleus along the lateral border of the ventral horn into a ventral direction (see C3 in Fig. 4). Traced further caudally into the rostral half of C_4, the lateral subnucleus is found at the ventrolateral margin of the ventral horn concomitantly with the total disappearance of the ventral edge of the ventral horn (see Fig. 3h and R4 in Fig. 4). This ventral position of the subnucleus is retained in caudal C_4 levels, but the neurons become progressively detached from the lateral margin of the gray matter (see Fig. 3i and C4 in Fig. 4).

This medial shifting of the lateral subnucleus concurs with the appearance of the lateral cell groups in the brachial plexus enlargement, being evident at levels from rostral C_5 to caudal C_7 (see Fig. 3j–o and R5–C7 in Fig. 4). Therefore, in the cervical intumescence, the lateral subnucleus of the spinal XI nuclear complex occupies an intermediate ventral position between the ventromedial and ventrolateral cell groups. Below mid-C_5 the diameter and the neuronal density of the lateral subnucleus decrease to taper in the caudal half of the 6th cervical segment (see Fig. 3k–m and C5–C6 in Fig. 4). Labeled neurons of the spinal XI nuclear complex are scattered throughout the 7th cervical segment (see Fig. 3n, o and R7 and C7 in Fig. 4), but below the caudal boundary of this segment no labeled neurons are found.

Thus, it is apparent from serial transverse sections of the cervical cord that the spinal XI nuclear complex comprises two subnuclei which overlap longitudinally and occupy different mediolateral positions within the ventral horn (Figs. 3, 4). They are present as separate units, except in the 3rd cervical segment, where the tapering end of the medial subnucleus appears to fuse with the lateral subnucleus at the latter's medial boundary (see Fig. 3e, f and R3 and C3 in Fig. 4).

Serial frontal sections of the cervical cord (following HRP exposure of the spinal XI nerve proximal to its ramification) reveal another outstanding feature of the spinal XI nuclear complex, i.e., its constant position within the ventral gray column (Fig. 5a, b). Although the positions of both subnuclei appear to shift along their rostrocaudal extents, it is evident that the longitudinal axis of either of the subnuclei forms a more or less straight line. Throughout their extent, the positions of both subnuclei within the gray matter as well as relative to each other are practically identical (see Fig. 4). Their displacement is only apparent and is caused by changes in cell grouping about them which alter the configuration of the ventral horn. Thus, additional neuronal groups, present at the dorsomedial border of the ventral horn below rostral C_1, appear to dislodge the medial subnucleus from a marginal into a more internal position (Fig. 5a). On the other hand, shortening of the ventral tip of the ventral horn, i.e., the progressive reduction of the ventral neuronal group below mid-C_3, causes a relative displacement of the lateral subnucleus into a ventral direction within the gray matter (see Fig. 3f–h and C3 and R4 in Fig. 4). Below mid-C_4, the concurrent lateral extension of the ventral horn accommodating brachial plexus neurons appears to push the lateral subnucleus from its ventrolateral position medialward (cf. Fig. 3i–o, C4–C7 in Fig. 4, and Fig. 5b).

By superimposing camera lucida drawings of the labeled spinal XI nuclear complex in the total number of representative transverse sections (14) with reference to a sagittal plane through the center of the central canal (see Fig. 4), additional and strikingly geometric features of the nuclear position are found.

The long axes of the subnuclei of the spinal XI nuclear complex lie parallel, both to the longitudinal axis of the central canal and to each other, throughout their rostrocaudal extents (Figs. 4, 6). Furthermore, the longitudinal axis of the lateral subnucleus is found to lie at a constant distance from the central canal on a virtual plane that is inclined at a fixed acute angle (approximately 60°) toward the median sagittal plane (Fig. 6). The long axis of the medial subnucleus is also found at a constant but shorter distance from the central canal on a virtual plane that is inclined at a smaller but also fixed angle toward the median sagittal plane (Fig. 6). In relation to that virtual radial plane inter-

15

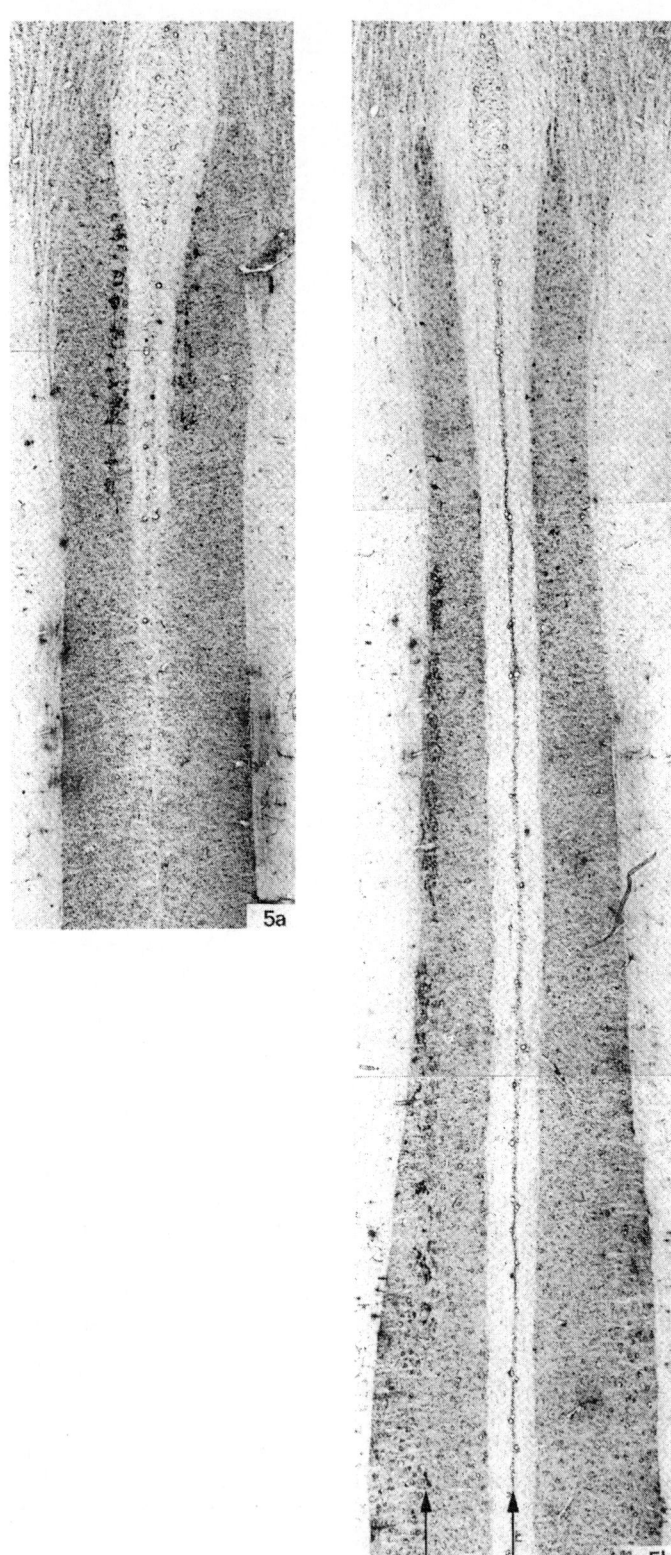

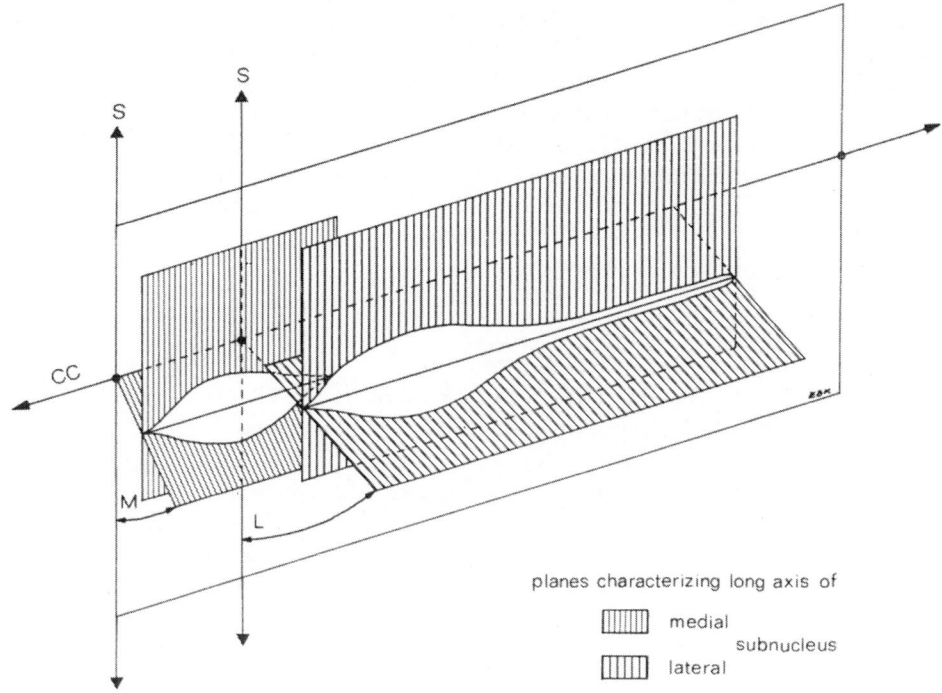

planes characterizing long axis of

medial
subnucleus
lateral

Fig. 6. Simplified stereodiagram of the spinal accessory (XI) nuclear complex in a lateral view to illustrate the geometric constancy of the nuclear position. A pair of intersecting virtual planes is used to characterize the position of each of the subnuclei, the line of intersection indicating the long axis of the subnucleus. The long axis of the lateral subnucleus is found at a constant and characteristic distance from the longitudinal axis of the central canal (*CC*) on a virtual radial plane, forming a constant and characteristic angle (*L*) with the median sagittal plane (*S*). Compared with the long axis of the lateral subnucleus, the axis of its medial fellow is characterized by the line of intersection of a radial plane, inclined at a smaller angle (*M*) toward the median sagittal plane, with a parasagittal plane, which lies more medial in the cervical cord (see Sect. 3.3 and 4.1.)

secting the long axes of both the central canal and the lateral subnucleus, the medial subnucleus lies slightly ventral, with its dorsolateral boundary touching the plane. With reference to the generally used frontal planes, however, the medial subnucleus lies dorsal to its lateral fellow.

Both subnuclei of the spinal XI nuclear complex are spindle-shaped. The spinal XI motoneurons tend to be more tightly packed at the intermediate

◁ **Fig. 5a, b.** Composite photomicrographs of frontal sections through the cervical cord, showing the medial (**a**) and lateral (**b**) subnucleus of the spinal accessory (XI) nuclear complex labeled with HRP following bilateral (**a**) and unilateral (**b**) exposure of the spinal XI nerve proximal to its ramification. **a** Note that the longitudinal axis of the medial subnucleus forms a more or less straight line in the ventral gray column. Additional neuronal groups at the dorsomedial border of the ventral horn, present below rostral levels of C_1, appear to dislodge the medial subnucleus from its marginal location into a more internal position. × 20. **b** Note the constant distance of the long axis of the lateral subnucleus from the ventral median fissure (*arrows*). The lateral extension of the ventral horn at levels of the cervical intumescence causes only an apparent displacement of the lateral subnucleus into a medial direction; × 20

17

portions of the subnuclei with larger diameters, while in the tapering rostral and caudal extremities of the subnuclei the neurons are more loosely arranged. In addition, a decrease in cellular density is found at more or less regular intervals (Fig. 5a, b). The neuronal density often varies abruptly from section to section. This periodically fluctuating density of labeled neurons is found predominantly in the lateral subnucleus, being most prominent in serial sagittal sections. Thus, the lateral subnucleus gives the appearance of a spindle-shaped structure with densely clustered motoneurons within its core and with periodically added neurons on its periphery.

3.4 Emergent Course of Axons Arising in the Spinal Accessory (XI) Nuclear Complex and the Intramedullary Ascending Spinal XI Fasciculus

The enhanced sensitivity of the tetramethyl benzidine procedure for HRP histochemistry (Mesulam 1976, 1978) makes possible the visualization of retrogradely transported HRP within axons. Advantage has been taken of this technique to examine the intramedullary course of the spinal XI axons (labeled after application of HRP to the spinal XI nerve proximal to its ramification) in serial transverse, frontal, and sagittal sections.

Spinal XI axons arising from the caudal part of the lateral subnucleus in the 7th and 6th cervical segments do not emerge at the level of their perikarya. After traversing the ventral horn, they curve posteriorly along the dorsolateral surface of the ventral horn toward a region midway between the ventral and dorsal horns. There, these nerve fibers bend rostrally to ascend as a distinct bundle through the reticular formation (Fig. 7). This bundle of caudal spinal XI axons forms the origin of an intramedullary ascending spinal XI fasciculus, constantly receiving axons from the spinal XI nucleus as it ascends throughout the cervical cord.

The caudalmost rootlets of the spinal XI nerve are seen to make their exit from the spinal cord at caudal levels of the 5th cervical segment (see C5 in Fig. 7). These spinal XI rootlets and several others emerging at any particular level above caudal C_5 arise from different sources and pursue a different intramedullary course. The majority of the emerging rootlets are spinal XI fibers which leave the intramedullary ascending fasciculus by curving laterally to traverse the lateral funiculus (Fig. 8a) along the ventrolateral border of the dorsal horn. These rootlets make their exit from the spinal cord at a short distance ventral to the entrance zone of the dorsal spinal roots (Fig. 7). Other spinal XI fibers emerge from the spinal cord at the level of their perikarya by following an indirect, double-bended route (Fig. 7). Arising from the spinal XI nucleus, these axons pass almost directly dorsolateralward through the ventral horn. Upon reaching the lateral boundary of the ventral horn, these fibers take a dorsomedial course along the dorsal aspect of the ventral horn (Figs. 7, 8b). Close and ventrolateral to the ascending spinal XI fasciculus, these rootlets bend sharply laterally to exit from the spinal cord slightly ventral to those issuing from the ascending fasciculus (Figs. 7, 8c).

Some of the spinal XI fibers, on reaching the surface of the cord, ascend close under the pia mater for a short distance before leaving the cord. The

18

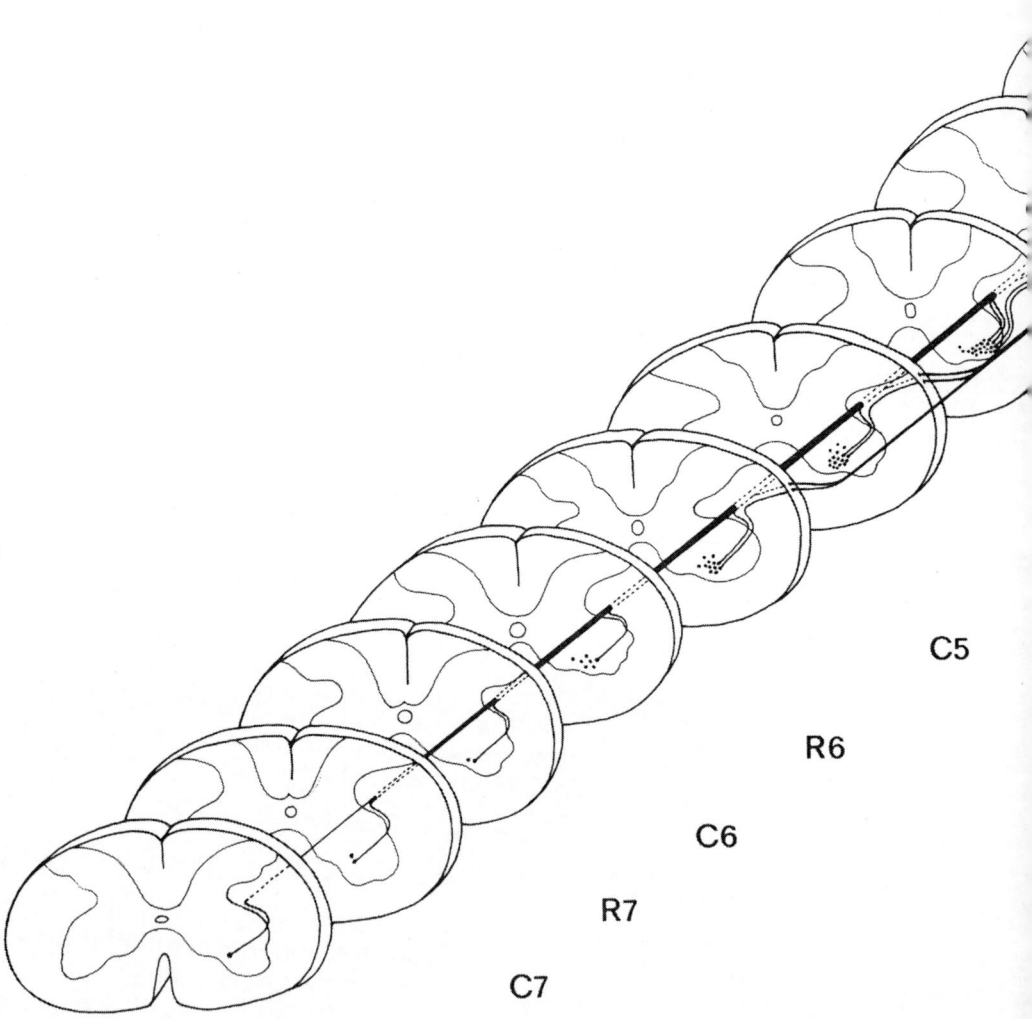

C5

R6

C6

R7

C7

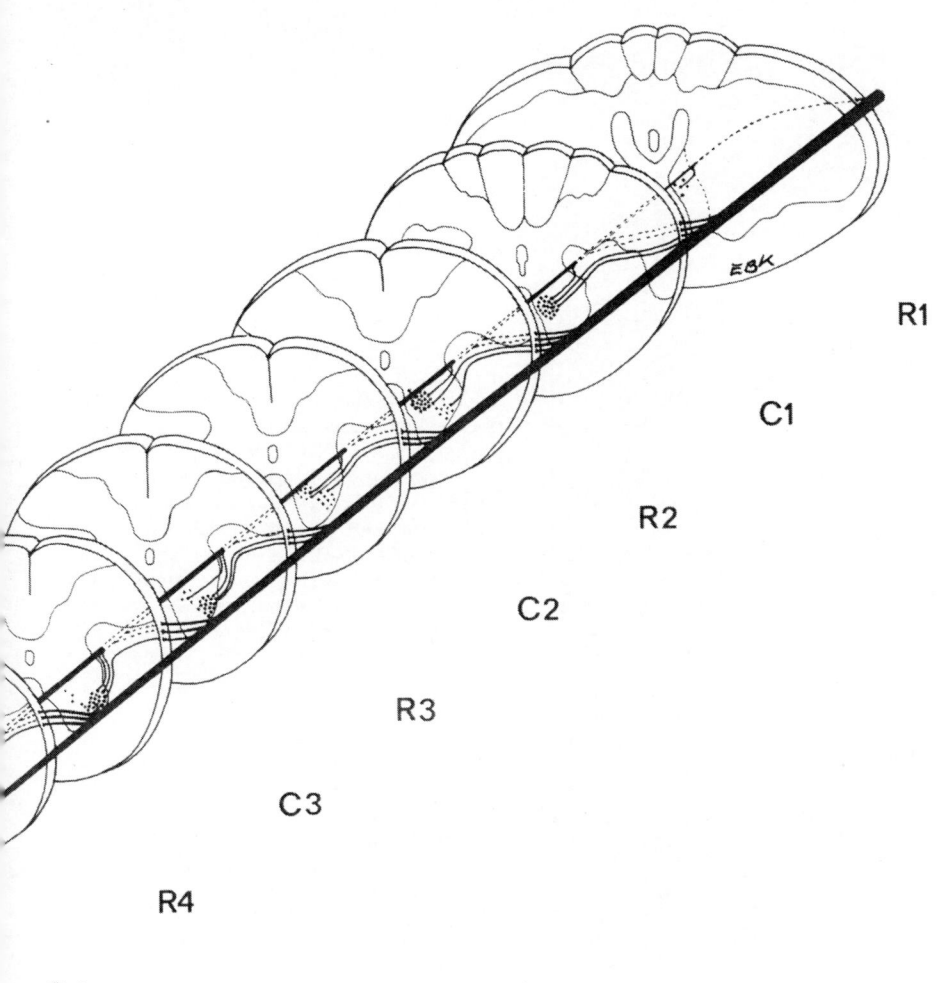

R1

C1

R2

C2

R3

C3

R4

C4

ᵧ. 7. Projection drawings of 14 representative transverse sections through the cervical cord at rostral
) and caudal (*C*) levels of the cervical segments $C_1(1)$–$C_7(7)$. The serial reconstructions are arranged
a perspective view from caudal (*bottom left*) to rostral to show the indirect emergent course taken
 the spinal accessory (XI) axons. Note that nerve fibers arising in the caudal portion of the spinal XI
clear complex (C7–R6) do not emerge at these levels but form the origin of an intramedullary as-
ading spinal XI fasciculus. The caudalmost spinal XI rootlets make their exit from the spinal cord
 caudal levels of C_5 (C5), thereby forming the origin of the extramedullary ascending spinal XI
ink. Note also that the spinal XI axons emerging at the level of their perikarya follow an indirect
uble-bended route to exit from the spinal cord slightly ventral to those issuing from the intramedul-
y ascending spinal XI fasciculus (Sect. 3.4; for a more detailed description and compare with Fig. 8)

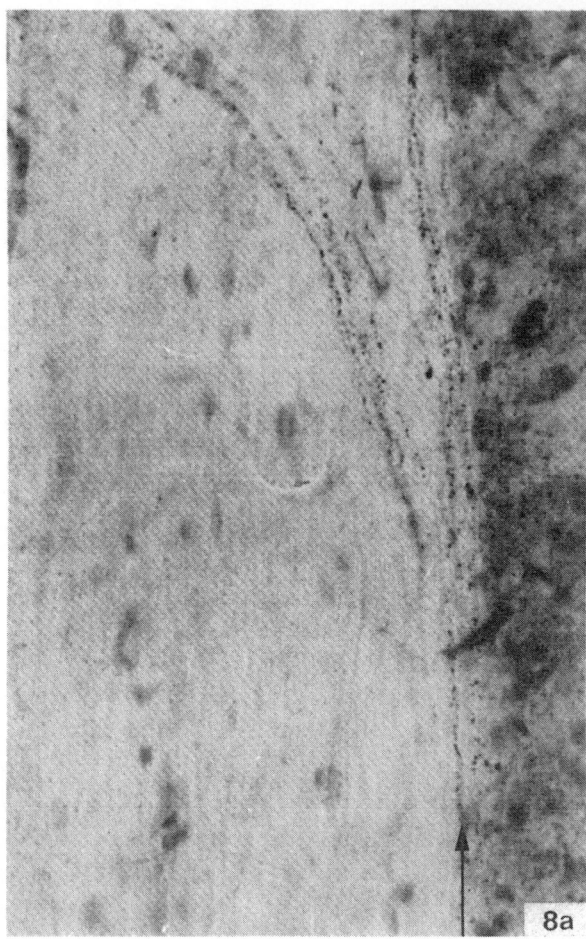

Fig. 8a–c. Photomicrographs illustrating the indirect emergent course of spinal accessory (XI) axons retrogradely labeled with HRP. **a** Frontal section through the cervical cord at the base of the ventral gray column (*right*). Note the bundle of nerve fibers which leave the intramedullary ascending spinal XI fasciculus (↗) by curving laterally to traverse the lateral funiculus; × 500. **b** Transverse section through the spinal cord at the junction of the 3rd and 4th cervical segments. Axons arising in the lateral subnucleus of the spinal XI nuclear complex take a dorsal course along the dorsolateral aspect of the ventral horn toward a region midway between ventral and dorsal horns. There (↗), these fibers bend sharply laterally to traverse the lateral funiculus; × 90. **c** Transverse section through the spinal cord at the junction of the 2nd and 3rd cervical segments. Note that spinal XI axons emerging at the level of their perikarya (↗) exit from the spinal cord slightly ventral to those issuing from the spinal XI fasciculus (↘); × 90

8a

axons ascending intramedullary in the spinal XI fasciculus often form small bundles which are made up of overlapping nerve fibers derived from different cervical levels. Bundles arising at lower levels overlap other bundles which originate at higher cervical levels. Owing to the mixing of the nerve fibers in the spinal XI fasciculus, the distance over which individual axons ascend before making their exit is difficult to determine. Nevertheless, it is evident that spinal XI axons emerge either at the level of their origin or at levels and in segments rostral to their perikarya. In other words, the perikarya of spinal XI rootlets which emerge at a particular spinal level lie either at this level or at levels and in segments always below their exit. Although the precise level of their origin is uncertain, it seems highly probable that those spinal XI axons joining the intramedullary fasciculus ascend for one or even more cervical segments before making their exit from the spinal cord. This follows from the fact that the spinal XI axons arising in C_7 emerge at least two segments above.

In addition to these differences in the intramedullary courses of the emerging rootlets, as described above, it should be noted at this point that the emergent routes of the spinal XI axons are progressively shifted into a dorsal direction in rostral cervical segments. Thus, the farther the spinal XI fibers are traced

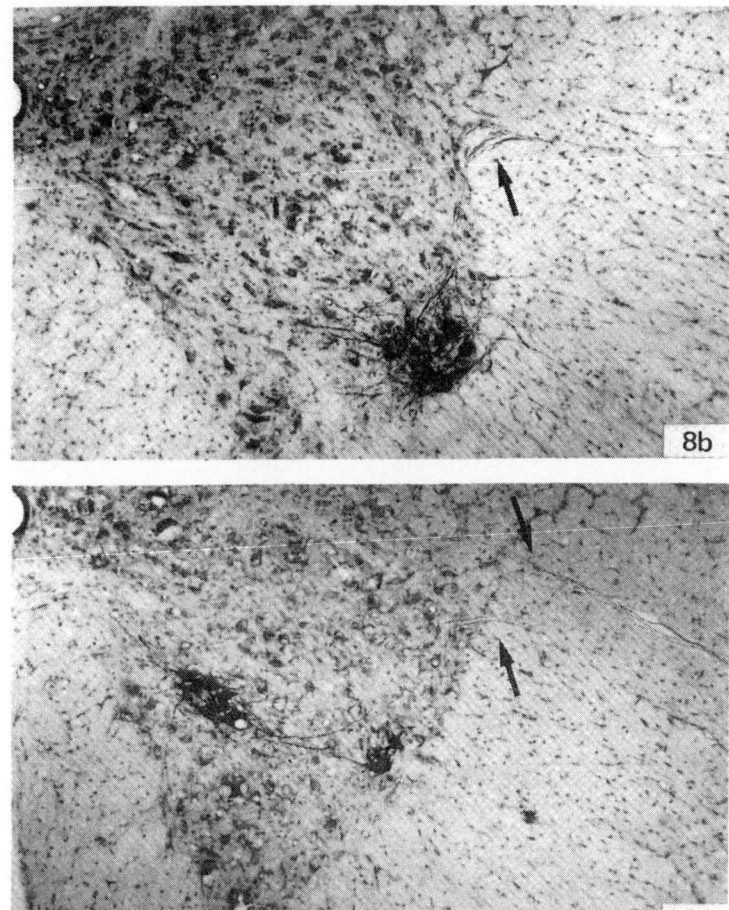

Fig. 8b, c.
Legend see p. 20

rostralward in the cervical cord, the more they approach the dorsal horn and the substantia gelatinosa during their intramedullary course, as well as the entrance zone of the dorsal spinal roots at their exits (Fig. 7). However, this dorsal displacement of the spinal XI rootlets appears to be only relative to the progressive enlargement and ventral extension of the dorsal horn and the substantia gelatinosa at rostral cervical levels.

After their emergence in the caudal half of the 5th cervical segment, the caudalmost rootlets incline rostrally, thereby constituting the origin of the extramedullary ascending spinal XI trunk. As higher rootlets are added, the trunk increases in size as it ascends, being situated between the dorsal roots and the ligamentum denticulatum and medial to the dorsal root ganglia. Upon reaching caudal levels of the medulla oblongata, the spinal XI trunk curves ventrally to join the vagoaccessory trunk.

Although the spinal XI trunk is intimately related to the dorsal roots and dorsal root ganglia of the upper cervical nerves, no anastomotic branches and intermingling of their fiber components are found in the rat. Occasionally, scattered small ganglion cells are found in the spinal XI trunk along its intravertebral course.

3.5 Somatotopic Organization of the Spinal Accessory (XI) Nucleus

After HRP exposure of the TRAP nerve beyond its junction with the communicating branch of the cervical plexus, but proximal to the entrance of its muscular branches into the TRAP muscle (at site 2 in Fig. 1), labeled motoneurons are found exclusively in the lateral subnucleus of the spinal XI nuclear complex. Within this subnucleus, TRAP motoneurons are present in the region from mid-C_2 to caudal levels of C_7 (Fig. 9). The largest number of labeled TRAP neurons is found in the region from rostral levels of C_3 to the junction of cervical segments C_4 and C_5, concurrently with the maximally increased diameter of the lateral subnucleus at these levels.

From rostral levels of C_5 to mid-C_6, the number of TRAP neurons decreases concomitantly with the neuronal density of the lateral subnucleus. Labeled TRAP neurons become more and more loosely arranged from caudal C_6 to caudal C_7, at levels where the lateral subnucleus is found to taper. Consequently,

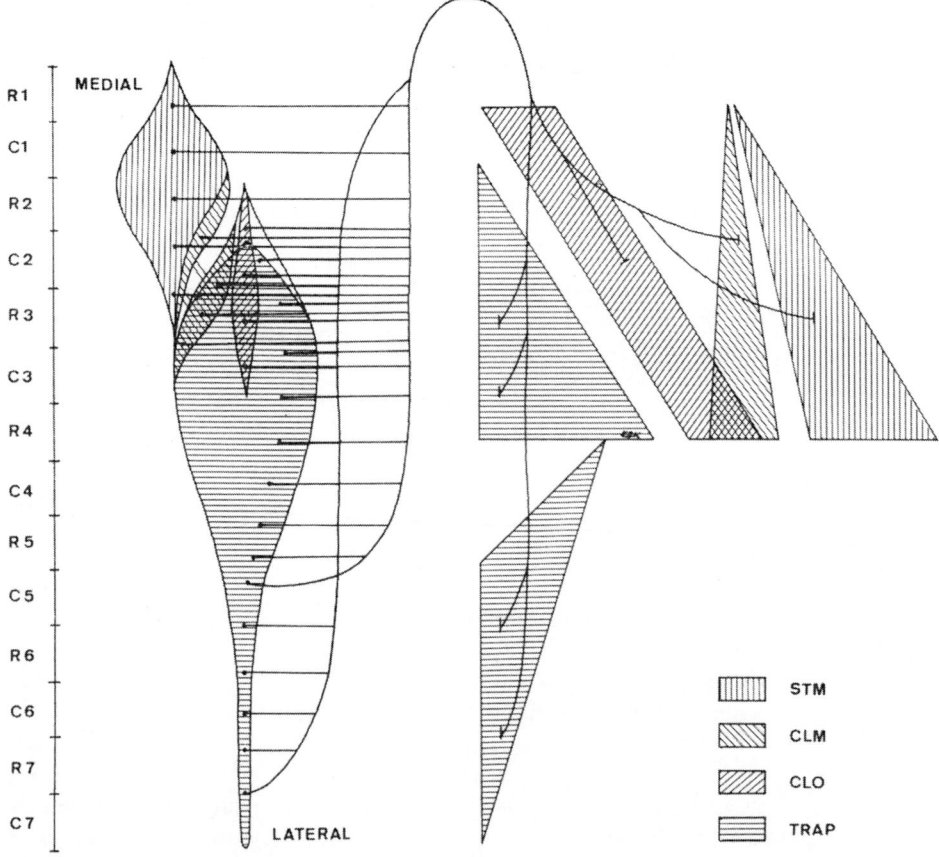

Fig. 9. Somatotopic organization of the spinal accessory (XI) nuclear complex in the rat: a summary scheme illustrating the arrangement of the motoneuronal groups innervating four spinal XI muscles within the medial and lateral subnucleus. The information presented is compounded from several labeling experiments (see Chap. 2 for further details) and projected diagramatically upon an oblique frontal section through the spinal XI nuclear complex from rostral (*R*) levels of $C_1(1)$ to caudal (*C*) levels of $C_7(7)$ (see Sects. 3.5 and 4.2.2 for a detailed description)

22

the TRAP motoneurons appear to constitute the lateral subnucleus of the spinal XI nuclear complex, at least from caudal levels of C_3 to the caudal extremity of this subnucleus.

Axons of the CLO nerve, exposed to HRP proximal to their entrance into the corresponding muscle (at site 3 in Fig. 1), transported the enzyme retrogradely to only a small number of motoneurons within the lateral subnucleus of the spinal XI nuclear complex. Labeled perikarya are observed exclusively in the rostral part of this subnucleus, from the junction of cervical segments C_1 and C_2 to that of C_3 and C_4. Motoneurons innervating the CLO are most frequent at adjacent levels of caudal C_2 and rostral C_3 (Fig. 9).

Following exposure of the CLM nerve, just proximal to its entrance into the homonymous muscle (at site 4 in Fig. 1), labeled motoneurons are found in both the medial and lateral subnucleus of the spinal XI nuclear complex. Whereas only a few CLM motoneurons are labeled within the medial subnucleus, a larger number are found in its lateral fellow. The medial CLM neurons, extending from caudal levels of C_1 to rostral levels of C_3, are most frequently found in mid-C_2 of the medial subnucleus. CLM neurons labeled within the lateral subnucleus are located from rostral levels of C_2 to caudal levels of C_3 and are most common from mid-C_2 to rostral C_3 (Fig. 9).

Although the CLM neurons lying in the lateral subnucleus are found at levels corresponding to those of the CLO neurons, the latter are outnumbered by the CLM neurons. Moreover, the maximal distribution of the CLM neurons starts in the lateral subnucleus at more rostral levels and exceeds that of the CLO neurons in all directions.

The neurons innervating the CLM muscle appear to be situated adjacent to the lateral rim of the medial subnucleus and in the medial part of the lateral subnucleus, as well as in that region where the facing margins of the subnuclei merge into each other (Fig. 9). While the CLM neurons are present in the more caudal part of the medial subnucleus, they lie in the rostral extremity of the lateral subnucleus. However, the longitudinal extension of the lateral group is slightly shifted in a caudal direction in relation to the medial.

Motoneurons, labeled after exposure of the STM nerve proximal to its entrance into the muscular hilus (at site 5 in Fig. 1), are confined to the medial subnucleus. They are found throughout the entire rostrocaudal extent of this subnucleus, i.e., from the medullospinal transition to mid-C_3 spinal levels (Fig. 9). Labeled neurons are most abundant between mid-C_1 and mid-C_2. The number of STM neurons gradually diminishes between caudal C_2 and rostral C_3, concomitantly with the neuronal density of the medial subnucleus. Thus, the neurons of the STM and those of the medial subnucleus show similar maxima of distribution and identical rostrocaudal extents.

3.6 Communicating Branches of the Cervical Plexus

Although several muscular branches of the spinal XI nerve were exposed to HRP distal to their union with the communicating branches of the cervical plexus, no labeled motoneurons were identified outside the spinal XI nuclear complex. It is thus highly probable that the spinal XI muscles receive their entire motor innervation from the spinal XI nerve.

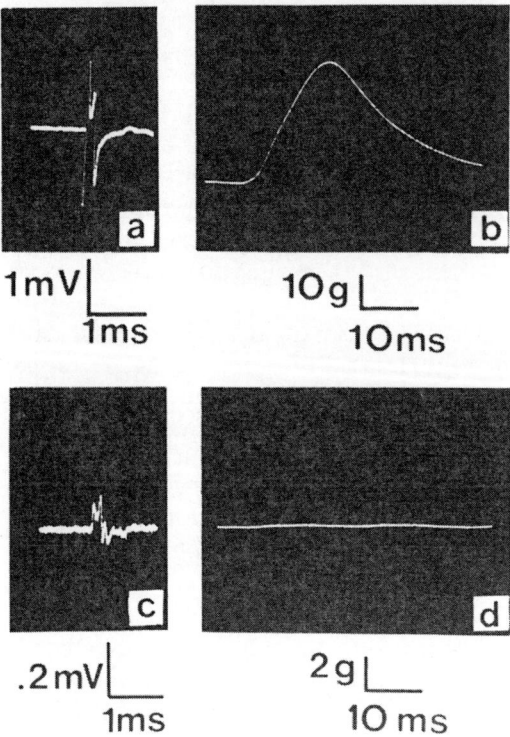

Fig. 10 a–d. Action potentials of the TRAP nerve, elicited by stimulation of **a** the spinal accessory (XI) nerve and **c** the communicating branch of the cervical plexus. Mechanograms of the superior TRAP recorded upon stimulation of **b** the spinal XI nerve and **d** the communicating branch of the cervical plexus

An additional attempt, using the cut-nerve exposure to HRP, was made to clarify the conflicting data, namely, the question of whether the communicating branches of the cervical plexus supply motor fibers to parts of the spinal XI muscles or are proprioceptive nerves in the rat. In a complementary experimental series HRP was applied to the cut ends of the muscular branches of the spinal XI nerve proximal to their communication with cervical rami, and, secondly, to the cut ends of these communicating branches proximal to their union with the muscular rami of the spinal XI nerve.

Following HRP exposure of the TRAP ramus of the spinal XI nerve proximal to its union with the communicating branch (at site 6 in Fig. 1), the same number of labeled neurons is found within the lateral subnucleus in corresponding transverse and longitudinal directions as after exposure of the TRAP nerve below its junction with the cervical ramus. On the other hand, HRP exposure of the cervical ramus communicating with the TRAP ramus of the spinal XI nerve (at site 8 in Fig. 1) resulted in labeling of spinal ganglion cells in the ganglia of the 3rd–5th cervical nerves, while no labeled motoneurons were found throughout the entire cervical cord.

The axons of the common trunk of CLM and STM rami, exposed proximal to their communication with branches of the cervical plexus (at site 7 in Fig. 1), transported HRP retrogradely to motoneurons of both subnuclei of the spinal

XI nuclear complex. Labeled neurons, found in the rostromedial part of the lateral subnucleus, confirm the presence of CLM motoneurons within this subnucleus. In addition, the motoneurons of the entire medial subnucleus appear to be labeled. Compared with the labeling after HRP exposure of the STM nerve, an increased number of labeled motoneurons is found in the medial subnucleus after these experiments. This increase in labeled neurons, especially prominent throughout C_2 and at rostral levels of C_3, is consistent with the presence of some CLM motoneurons at all levels of the medial subnucleus, except in its rostral portion.

Application of HRP to the cut ends of several cervical branches communicating with the CLM and STM rami (at site 9 in Fig. 1) resulted in labeling of spinal ganglion cells in the ganglia of the 3rd–5th cervical nerves. No labeled motoneurons are seen throughout the cervical cord, neither within nor outside the spinal XI nuclear complex.

It seems, therefore, most likely from these experiments that in the rat the spinal XI musculature receives no motor fibers from the cervical ventral rami. Rather, it has to be concluded that these muscles receive their entire motor supply from the spinal XI nerve. On the other hand, the communicating branches of the cervical plexus join the muscular branches of the spinal XI nerve to supply proprioceptive fibers to the spinal XI muscles. The latter fibers arise in spinal ganglion cells of the 3rd–5th cervical root ganglia.

In order to confirm these results, some additional stimulation experiments were performed. Stimulation of the spinal XI nerve or its TRAP ramus resulted in strong contractions of the superior TRAP. Accordingly, action potentials of high amplitude were recorded from the TRAP nerve (Fig. 10a, b). On the other hand, stimulation of the cervical branch communicating with the TRAP ramus produced low-amplitude action potentials in the TRAP nerve and no contractions of the superior TRAP (Fig. 10c, d).

One week after transsection of the spinal XI nerve, stimulation of neither the TRAP nerve nor the communicating branch of the cervical plexus caused contractions of the superior TRAP.

4 Discussion

4.1 Topographic and Geometric Position of the Spinal Accessory (XI) Nuclear Complex

There have been many more or less successful attempts to localize the spinal XI nucleus in the cervical cord of various species, including man. The localization of this nucleus is, however, based mainly on structural data (Henle 1871; Darkschewitsch 1885; Kaiser 1891; Koelliker 1896; Lubosch 1899; Gehuchten 1900; Bruce 1901; Hepburn and Waterston 1904; Angulo y González 1927; Hogg 1928; Pearson 1938; Kimmel 1940; Yoda 1940; Romanes 1941; Rexed 1952, 1954; Pearson et al. 1964; Matsushita 1970; Silver and Wolstencroft 1971), corroborated only to some degree by subsequent experimental studies (Straus and Howell 1936; Holomáňová et al. 1972; Gura and Limanskii 1977; Rapoport 1978; Gottschall et al. 1980b; Robards et al. 1980; Karim and Hoo Nah 1981). Therefore, it is not surprising that the spinal XI nucleus of mammals has been localized with such a variety of extensions and positions within the ventral gray column.

In finding the spinal XI nuclear complex to extend from the medullospinal transition to caudal levels of C_7, our results differ from those of most previous investigations. Compared with the rostrocaudal extent of the spinal XI nucleus found in the present investigation, the previously reported extents appear too contracted. Although the longitudinal expansion of this nucleus had been underestimated, the spinal XI nucleus had always been identified in cervical segments corresponding to those levels (C_2–C_4) at which we find both subnuclei to be present or either of them maximally developed. Therefore, it seems most likely that there are difficulties in localizing the rostral and caudal portions of the spinal XI nucleus by solely descriptive anatomy. The assumption that the spinal XI nucleus terminates in the 4th or 5th cervical segment (Streeter 1905; Rexed 1954; Holomáňová et al. 1972; Gura and Limanskii 1977; Rapoport 1978) is probably due to the gradual decrease of the nucleus below this level. Furthermore, this caudal limit of the nucleus may have been extrapolated from the emergence of the caudalmost rootlets at this level in assuming the exit of the spinal XI rootlets to be at the level of their origin. The rostral part of the spinal XI nuclear complex is also difficult to identify without concomitant evaluation by experimental procedures. Although a number of authors assumed a dorsomedial position of the rostral part of the spinal XI nucleus (Henle 1871; Kaiser 1891; Bruce 1901; Hepburn and Waterston 1904; Angulo y González 1927; Hogg 1928; Yoda 1940; Romanes 1941; Matsushita 1970; Holomáňová et al. 1972, 1973; Gura and Limanskii 1977; Rapoport 1978; Gottschall et al. 1980b; Karim and Hoo Nah 1981), most cytoarchitectonic studies followed

Rexed (1952, 1954) in referring the dorsomedial cell groups in the upper cervical segments to the "commissural" nucleus (J. v. Lenhossék 1855; M. v. Lenhossék 1889, 1895; Ramón y Cajal 1909) and "centrodorsal" nucleus. In consequence, this region in the ventral horn of the cervical cord is thought to be propriospinal and to contain no motoneurons. This easily explains why the spinal XI nucleus had been assumed to lie in a lateral position within the ventral horn throughout its extent, even in the 1st cervical segment (e.g., Rexed 1952, 1954).

The presence of spinal XI motoneurons in a medial position at rostral cervical levels and in a lateral position at more caudal levels had been reported by only a few authors. Whereas most of them had assumed there to be a positional shift of a continuous cell column from medial in rostral segments to lateral in more caudal cervical segments (Kaiser 1891; Bruce 1901; Yoda 1940; Romanes 1941; Matsushita 1970; Silver and Wolstencroft 1971; Holomáňová et al. 1972, 1973; Gura and Limanskii 1977), the evidence for the existence of two distinct entities of the spinal XI nucleus is hardly sufficient (Hepburn and Waterston 1904; Angulo y González 1927; Rapoport 1978).

Our results leave no doubt that the spinal XI nucleus comprises two distinct subnuclei which overlap in a longitudinal direction, occupying at the same time different positions within the ventral horn. However, the facing borders of the subnuclei appear to fuse together at levels where the medial subnucleus terminates and its lateral fellow is maximally developed. This apparent fusion of the subnuclei in the 3rd cervical segment cautions us against regarding the spinal XI nuclear complex as consisting of two individual motor columns. Rather, we are inclined to assume that the rostral part of a common column of spinal XI motoneurons has been split during its development into two functionally different components. This might well have occurred, in accordance with the concept of neurobiotaxis (Ariëns Kappers 1914, 1927; Ariëns Kappers et al. 1936), through migration of the neurons of the orimentary medial subnucleus toward the sources of their most important stimulation.

It remains to be seen whether our conceptions of a common spinal XI column and of a migration of rostral parts into a medial position is correct. In the meantime, we may take them merely as a suggestion proving that a rational explanation is at least conceivable.

However, some tentative idea concerning the splitting of a common spinal XI column can be had from the particular position of the spinal XI nuclear complex within the ventral horn, which may be characterized by using geometric principles.

It is widely held that the spinal XI subnuclei, like other motor nuclei in the cervical cord (e.g., Rapoport 1978; Gottschall et al. 1980b; Gottschall 1981; Karim and Hoo Nah 1981), change their position throughout their rostrocaudal extents. However, the spinal XI subnuclei appear to change merely their topographic position in the ventral horn, i.e., their spatial relation to other neuronal groups. This apparent positional shift is most striking in the lateral subnucleus on account of its longitudinal expansion throughout regions of the cervical cord, where considerable rearrangements in the cell grouping alter the configuration of the ventral horn. Thus, the reduction of the ventral tip motoneurons innervating dorsal neck muscles (Richmond et al. 1978) at caudal levels of C_3 causes an apparent ventral shift of the lateral subnucleus, whereas its medial displacement below C_4 is related to the development of the cervical intumescence.

27

In consequence, we found that the geometric position of the spinal XI nucleus is much the same throughout its extent, with the long axes of the subnuclei forming almost straight lines within the cervical cord. The fact that the motor nuclei of the spinal cord change their position in the ventral horn only relative to other motoneuronal groups had been stressed by some investigators (Angulo y González 1927; Pearson 1938; Duron et al. 1979; Webber et al. 1979; Kuzuhara and Chou 1980; Rikard-Bell and Bystrzycka 1980). However, geometric principles which may be used to characterize the constant position of a particular nucleus have as yet not been found. Probably, on account of the considerable length of the lateral XI subnucleus, we could not escape the impression of the positional regularity of the spinal XI nuclear complex.

Thus, the geometric position of each of the subnuclei of the spinal XI nuclear complex in the cervical cord may be characterized (Fig. 6) by two intersecting planes, the line of intersection indicating the long axis of the subnucleus. One of these virtual planes, being parallel to the median sagittal plane through the central canal, defines the laterality of the nuclear position. Another plane, lying radially with regard to the long axis of the central canal, forms a constant and particular angle with the median sagittal plane. This imaginary radial plane gives information about the position of the nucleus in a ventrodorsal direction. The long axis of the subnucleus is found on this radial plane at a constant and particular distance from the central canal, this distance in turn being determined by the parasagittal plane. Therefore, the geometric constancy of the nuclear position may be characterized by the relationship between a constant angle and a constant distance, namely, the relation of a constant angle, lying between a radial plane and the median sagittal plane, to a constant distance, measured along the radial plane from the central canal. This constant distance/constant angle relation is discoverable in the position of the long axis of both subnuclei of the spinal XI nuclear complex. However, compared with the long axis of the lateral subnucleus, the axis of its medial fellow is found on the line of intersection of a radial plane, inclined at a smaller angle toward the median sagittal plane, with a sagittal plane, which lies more medial in the cervical cord. Accordingly, the distance/angle relation characterizing a nuclear long axis appears to be not only constant throughout the extent of a subnucleus but also peculiar to a subnucleus.

Since this striking geometric arrangement of motor nuclei has not been observed as yet, we analyzed camera lucida drawings of the cat phrenic nucleus (Webber et al. 1979) according to the above criteria. Surprisingly, the long axis of the phrenic nucleus is also found at a constant distance from the central canal, on a radial plane, which is inclined at a constant angle toward the median sagittal plane. As another species had been used in the study, we could not prove whether our assumption that the constant distance/constant angle relation is also peculiar to a motor nucleus was correct. However, it seems highly probable that the phrenic nucleus lies at a shorter distance from the central canal on a virtual radial plane with a smaller angle, as compared with the lateral subnucleus of the spinal XI nuclear complex.

Thus, it would be attractive to speculate that motor nuclei and motoneuronal groups in general are geometrically organized within the spinal cord. Moreover, the present observations suggest that the most prominent feature of this geometric order is an organization of the motoneuronal groups with respect to radial axes of the ventral horn. If developmental data are taken into account, the assumption of a radial organization of the motoneurons in the ventral horn seems to gain support. Thus, it is well known that neuroblasts arising from ependymal mitoses move along radial courses into the mantle layer where they differentiate (e.g., Angulo y González 1939; Weiss 1969).

With respect to the geometric arrangement of the spinal XI nuclear complex, another interesting feature becomes noticeable, i.e., although the spinal XI subnuclei change their position with reference to other neuronal groups, their spatial relation to each other is constant (Fig. 4). Therefore, the subnuclei appear to form a pair of neuronal columns which, although spatially separated, nevertheless maintain a fixed spatial relationship. This arrangement of the subnuclei in couples and their fusion at the caudal extremity of the medial subnucleus lend support to the idea that both subnuclei are derived from a common spinal XI column. We are therefore inclined to conclude that a (developmental?) factor has separated the motoneurons in the rostral portion of the spinal XI nucleus. In view of the fact that the lateral subnucleus retains its particular position from the 2nd to the 7th cervical segment, it seems a more likely assumption that part of the rostral spinal XI neurons have migrated into a medial rather than a lateral position.

To summarize, it seems reasonable to suggest that the particular position of a definite motor nucleus within the cervical cord may be characterized by a peculiar and constant radial geometry.

4.2 Somatotopic Organization of the Spinal Accessory (XI) Nuclear Complex: Functional Significance of the Spinal XI Muscles and of the Spinal XI Subnuclei

The present results clearly demonstrate that in the rat the spinal XI muscles receive their entire motor innervation from fibers arising in the spinal XI nuclear complex. No evidence of a motor component was found in the communicating branches of the cervical plexus. These branches of the cervical ventral rami, which anastomose extraspinally with muscular branches of the spinal XI nerve, contribute proprioceptive fibers to the spinal XI muscles of the rat. Our results are in agreement with those obtained in other species (Straus and Howell 1936; Corbin and Harrison 1938; Yee et al. 1939) and therefore support the concept of an exclusive motor supply to these muscles by the spinal XI nerve in mammals.

Compared with the work done on other cranial nerves, there have been only a few experimental approaches aimed at localizing the spinal XI motor nucleus. Moreover, even less is known about the arrangement of the motoneuronal groups innervating the individual muscles of the spinal XI musculature within this nucleus. Thus, the organization of the spinal XI nuclear complex has not been satisfactorily analyzed with the neuroanatomical methods currently available. There have been only a few attempts utilizing the technique of retrograde transport of HRP to localize motoneurons innervating particular spinal XI muscles within the cervical cord (Gottschall et al. 1980b; Robards et al. 1980; Karim and Hoo Nah 1981), and only one study on the somatotopic organization of the spinal XI nucleus (Rapoport 1978). However, our results concerning the somatotopic arrangement of the motoneurons within the spinal XI nucleus differ from those obtained by Rapoport (1978), this probably being partly because of the different techniques of HRP application. Injection of HRP directly into a muscle, a technique employed by Rapoport (1978), was found to produce spurious labeling of motoneurons resulting from diffusion of HRP to motor fibers of adjacent muscles (Richmond et al. 1978). On the other hand, it may be that the HRP solution does not sufficiently spread to all axons innervating a particular muscle. This problem arises especially in the case of large flat muscles, e.g., the TRAP, whose fiber bundles receive their motor supply stepwise via a series of filaments branching off from the TRAP nerve, which descends perpendicularly to the muscle fibers for a long distance. Exposure of the cut ends of individual motor nerves to HRP, a technique used in the present experiments, besides eliminating the problem of HRP spread to other nerves, insures an equalized exposure of all nerve fibers innervating a particular muscle (Richmond et al. 1978; Krammer et al. 1979).

Further inconsistencies appear to arise from the fact that deductions from human myology, although a convenient simplification in somatotopic studies, are not generally applicable to the anatomy and function of the spinal XI muscles in quadrupedal mammals. Thus, the spinal XI musculature of the cat and most mammals comprises not merely two muscles as assumed by Rapoport (1978) and Karim and Hoo Nah (1981), but four or five muscles (e.g., Streissler 1900; Schück 1913; Luther and Lubosch 1938; Jouffroy 1971), depending on whether superior and inferior TRAP are fused or not. Furthermore, owing to the absence of the clavicle in the cat (e.g., Ellenberger and Baum 1932; Sandstrom and Saltzman 1944), the muscles usually attached to it – the CLM and the CLO – are inserted into the forearm instead (e.g., Owen 1868; Ellenberger and Baum 1932). Therefore, in the cat, these muscles are primarily concerned with quadrupedal locomotion and cannot be regarded as functionally homologous to those in man.

4.2.1 Myology

The rat has a well-formed clavicular brace and, with the associated ligaments, possesses a shoulder mechanism comparable in this respect to that in man. Although structurally the shoulder mechanism of the rat resembles that of man (Sandstrom and Saltzman 1944), it exhibits a wider freedom of movement as it is adapted both to tractional movements, which require an adequate clavicular brace, and to quadrupedal locomotion, which is considerably facilitated by a relatively narrow shoulder width. The manner in which the shoulder can be adapted to the need of both movements, particularly as they are apt to occur conjunctly in the narrow passages used by rats, would seem to rest in the presence of the omosternum (Sandstrom and Saltzman 1944). By adding the required length to the clavicle, the omosternum (see e.g., Dawson 1925; Klima 1968; Klima 1973) allows the scapula to move from the thorax into the neck, thereby permitting a reduction in the shoulder width and increasing the versatility of the shoulder. In this extreme elevation of the shoulder girdle, the spinal XI muscles concerned are those which "descend" to their attachments into the clavicle and scapula. Thus, the superior TRAP, the CLO, and most probably the CLM muscles are prime movers of the shoulder girdle, effecting its forward movement into the neck region and the concomitant elongation of the clavicle. Since the superior TRAP in the rat is not attached to the skull (Streissler 1900) just as in many mammals (and sometimes in man), only three of the spinal XI muscles, namely, the STM, CLM, and CLO act directly on the atlanto-occipital and atlantoaxial joints. However, the movements produced by muscles acting on these joints cannot be deduced from human arthrology.

In contrast to species which hold their heads on an erect cervical column, in the rat the major axes of the vertebral column and skull are roughly in the same line (DuBrul 1950). In the former species the long axis of the skull is nearly at a 90° angle with that of the cervical column and, therefore, the planes of the foramen magnum and of the condyles face ventrally. On the contrary, in the rat the foramen magnum opens directly caudally (DuBrul 1950), and the body axis continues in the same general line. Thus, in those animals postured horizontally like the rat, the occipital condyles lie in a caudal instead of a ventral position. Accordingly, the movements of the head differ from those in man and in animals with erect or semierect posture. Thus, in man, rotatory movements of the head in the horizontal plane would correspond to turning movements in the rat. Rotatory movements in the rat, performed in a vertical plane, would correspond to turning movements in man (see e.g., Podivinský 1968). Furthermore, because of the caudal position of the occipital condyles, the STM, CLM, and CLO muscles, being attached rostral to the atlanto-occipital joints, pass ventral to the transverse axis of these joints and, therefore, jointly produce flexion. Since an occipital attachment of the superior TRAP is absent, none of the spinal XI muscles extends the head of the rat.

In quadrupedal mammals the ligamentum nuchae is an important factor suspending the head and modifying its flexion, thus functioning in association with the extensors of the head and neck (Jouffroy 1968). Among the extensors, the biventer and the complexus appear to be involved in the tonic activity needed to maintain a head position, whereas the splenius is used in rapid turning and rotatory movements (Abrahams and Rancier 1973; Richmond and Abrahams 1975; Abrahams 1977) jointly with the STM and CLM muscles. Among the three muscles of the spinal XI group attached to the skull, the STM seems to act primarily from below (see e.g., Owen 1868; Ellenberger and Baum 1932; Jouffroy 1971) and to be involved mainly in direction-specific movements of head and eyes.

4.2.2 Somatotopic Grouping of the Motoneurons

The foregoing survey of the particular anatomy of the spinal XI muscles and their main function in the rat must be taken into consideration if the term "functional interpretation" in respect of the somatotopic organization of a motor nucleus is to become more than a vague phrase. Evidence from descriptive and experimental anatomy, clinical pathology, and physiology has given rise to many theories as to the functional nature of the localizations of neuronal groups in the spinal motor system. Thus, it was concluded that the motoneuronal groups identifiable topographically within the ventral horn were not usually neuronal groups innervating individual muscles, but more often represented pools of neurons supplying muscles acting together, thereby producing a common effect on a particular joint (e.g., Goering 1928; Romanes 1964). Other investigators indicated that the cell groups in the ventral horn can be correlated equally well with particular peripheral nerves (see e.g., Reed 1940), limb or trunk segments (see e.g., Romanes 1941), and with muscles grouped on a morphological or developmental basis (see e.g., Angulo y González 1927). As we will try to show, there is some truth in most of these interpretations concerning the motoneuronal organization of the spinal XI nucleus, but none of them can be generally applied to the mutual relation of the motor cell groups to their target muscles.

Whereas discrete grouping of neurons innervating individual muscles tends to occur in the motor nuclei of cranial nerves (Krammer et al. 1979), in the spinal cord it seems to be only a local phenomenon, present in some areas but absent in others. Within the cervical cord only two individual motor nuclei are supposed to exist, i.e., the phrenic nucleus and the spinal XI nucleus. Thus, Rapoport (1978), on the basis of HRP studies, suggested that each of the two subnuclei of the spinal XI nucleus represents a motor column innervating an individual muscle, namely, the sternocleidomastoid and the trapezius.

However, the results of the present experiments necessitate a considerable modification of this concept of the motoneuronal organization in the spinal XI nuclear complex. In particular, the suggestion that each column of the spinal XI nucleus innervates one of the two spinal XI muscles is invalid, as neuronal groups innervating at least four muscles turned out to be pooled within two subnuclei. The motoneurons supplying two muscles – the STM and CLM – are found in the medial XI subnucleus, whereas the lateral XI subnucleus represents a motor pool for the innervation of at least three muscles. On the other hand, the motoneurons innervating an individual muscle are generally confined to only one of the subnuclei, with the exception of the CLM, whose neurons are present in both subnuclei. However, the CLM motoneurons are located in those parts of both subnuclei which, facing one another, fuse together at the caudal extremity of the medial subnucleus.

Although both subnuclei turned out to represent conjoined motor pools innervating at least two muscles, each of the subnuclei is dominated by motoneurons supplying a particular "leading" muscle. Thus, the neurons innervating the STM constitute the overwhelming majority of the perikarya within the medial subnucleus, being present at several levels of this subnucleus and forming both its extremities and its "belly". Compared with the great number of STM neurons, only a few motoneurons of the medial subnucleus innervate the CLM. These latter neurons are located in the lateral part of the medial subnucleus at all its levels except the rostral ones. The dominant neurons of the lateral subnucleus are obviously those innervating TRAP. However, these TRAP motoneurons are not found in the rostral part of this subnucleus, being absent above mid-C_2. Below this level the neurons supplying the TRAP are the leading ones and become the exclusive neurons of this subnucleus from caudal C_3 to caudal C_7. Thus, the TRAP neurons are those forming solely the bulging part and the caudal extremity of the lateral subnucleus. Within the rostral part of the lateral subnucleus, the neurons innervating two muscles – the CLO and CLM – are pooled with a considerable overlap. Both neuronal groups are present from the rostral extremity of the lateral subnucleus to caudal levels of C_3.

However, the neurons supplying the CLO are less numerous and have their maximum at slightly lower levels than those innervating the CLM. Within the lateral subnucleus, therefore, a craniocaudal sequence of motoneuronal groupings is indicated, whereby the neuronal groups overlap to a large extent in its rostral part. Whereas the neurons innervating the CLM and CLO have their maxima in the rostral part of the lateral subnucleus, the bulk of the TRAP neurons are identical with those of the subnucleus.

Rapoport (1978), however, found the neurons innervating the trapezius to be present from C_1 to C_5, having their largest concentration in the C_2–C_4 segments. On the basis of our observations (cf. Fig. 9), we assume that the axons innervating the CLO have been exposed to HRP in common with those of the superior TRAP, thus resulting in neuronal labeling at rostral levels of the lateral subnucleus. On the other hand, the absence of labeled motoneurons below C_5 may be due to failure of exposure of the axons innervating the inferior TRAP.

It is interesting to note that a beaded structure is present in the spinal XI nucleus, especially in its lateral subnucleus. This spatial rhythmic structure of the lateral subnucleus, appearing in more or less regular spacing, is most prominent in serial sagittal sections. This observation suggests that motoneurons are added periodically at the lateral and the medial surface of this subnucleus. Probably, this spatial rhythmicity of the lateral subnucleus is related to the maxima of the neurons innervating CLM, CLO, and TRAP which lie in rostrocaudal series and, further, to the peculiar nervous supply of the TRAP. Thus, its fiber bundles receive their motor supply via a rostrocaudal series of filaments branching off from the TRAP nerve. These myoneuronal units, supplied by successive filaments, may be related to the rhythmic swellings of the lateral subnucleus in its caudal half.

4.2.3 Significance of the Motoneuronal Pools

The question still remains: what can be the functional significance of the motor pools found within the spinal XI nuclear complex? Evidently, the spinal XI subnuclei do not represent motor columns innervating one muscle each, but pools of motoneurons supplying two or three muscles. If there is a pooling of motoneurons innervating muscles of identical function (e.g., Goering 1928; Romanes 1964), the presence of CLM, CLO, and TRAP motoneurons within the lateral subnucleus would suggest a common motoneuronal pool for elevators of the shoulder. However, neurons supplying the CLM are also present in the medial subnucleus, which in turn is dominated by neurons innervating the STM. On the assumption that the STM acts predominantly from below as a primary mover of head and neck, the presence of neurons innervating the CLM in the medial subnucleus may indicate that the CLM subserves head movements in addition to shoulder elevation.

Obviously, the spinal XI subnuclei do not represent motor pools innervating, respectively, flexors and extensors of the head, as proposed by Rapoport (1978). As outlined above, the STM, CLM, and CLO might flex head and neck – as suggested by geometric reasoning – but their motoneurons are localized in different subnuclei. Furthermore, the neurons innervating the potential flexors

– the CLO and CLM – lie in the lateral subnucleus, which in turn is dominated by neurons innervating TRAP, a muscle that does not act directly on the head. In view of the probable indirect extension of the head by the TRAP via its attachment to the ligamentum nuchae, a functional interpretation of the neuronal pools present in the spinal XI subnuclei with respect to innervation of flexors and extensors becomes even more impossible.

Likewise, the neuronal pools of each of the spinal XI subnuclei cannot be correlated with particular peripheral nerves (e.g., Marinesco 1898; Reed 1940), since the neurons arranged within the lateral subnucleus give rise to three different peripheral branches. However, the neurons pooled in the medial subnucleus tend to give rise to the ventral trunk of the spinal XI nerve which innervates the STM and CLM, while the dorsal division of the spinal XI nerve arises exclusively from neurons of the lateral subnucleus. On the other hand, the CLO nerve does branch off from the ventral division of the spinal XI nerve, although its neurons lie exclusively within the lateral subnucleus, which in turn gives rise to the dorsal division of the spinal XI nerve. Therefore, we conclude that the motoneuronal organization of the spinal XI nuclear complex is not related to peripheral branches of the spinal XI nerve.

4.2.3.1 Developmental Considerations

With reference to the theories on the significance of motor pools, we will examine whether the subnuclei of the spinal XI nuclear complex represent communities of neurons innervating muscles grouped on a developmental basis (Angulo y González 1927).

In the phylogenetic series, the trapezius muscle is said to make its first appearance in fishes, namely, in the chondrichthyes (see Addens 1933; Straus and Howell 1936; Luther and Lubosch 1938). Hence, it seems to occur during phylogeny concurrently with the first appearance of a cartilaginous shoulder girdle in this class of fishes. The trapezius is represented, e.g., in selachians, by a muscle which arises from the superficial fascia covering the epaxial muscles and is inserted into both the scapular homologue of the shoulder girdle and the epibranchial cartilage of the last gill arch (see Straus and Howell 1936; Luther and Lubosch 1938). Its attachment to the last epibranchial corresponds perfectly to that of the levatores arcuum branchialium to their respective epibranchials. The trapezius of these cartilaginous fishes elevates the shoulder girdle and the last gill arch, and is innervated, as far as is known, solely by a caudal twig of the series of vagus branches supplying the caudal part of the gill musculature (see Addens 1933; Straus and Howell 1936; Luther and Lubosch 1938).

This intimate relation of the trapezius to the branchial muscle plates in the phylogenesis of the muscle has been found to exist also during its ontogenetic development in a variety of vertebrates, from selachians to mammals (e.g., Edgeworth 1911, 1926). In spite of one isolated claim to the contrary (Favaro 1903), great importance should be accorded to Edgeworth's conclusions in view of the extensiveness of his investigations.

In amphibians the trapezius, though arising mainly from the superficial dorsal fascia, tends to shift its origin headward (Straus and Howell 1936; Luther and Lubosch 1938). Thus, in urodeles, two muscles can be identified – a dorsoscapularis, corresponding to a primitive trapezius, and a capitoscapularis. In anurans, in which only a capitoscapularis exists, this muscle more closely resembles a sternocleidomastoid than a trapezius. However, even in claviculated anurans, this head portion of the muscle is inserted into the scapula and, therefore, does not correspond to a muscle of the sternocleidomastoid group.

In reptiles, the accessory musculature extends its origin on the head in a ventral direction, and from the neck into the thoracic region (see Straus and Howell 1936; Luther and Lubosch 1938). The migration of the muscle in a caudal direction seems to occur concurrently with the formation of a neck in sauropsides. Thus, neck and neck muscles lengthen, the shoulder girdle drops, and at the same time the ability to rotate head and neck increases conspicuously. Moreover, with the first appearance of the secondary sternal elements in reptiles (see Luther and Lubosch 1938), the accessory musculature modifies its insertion, and the sternocleidomastoid muscles begin to differentiate. Hence, in sauropsides, two muscles are present – a capitosternalis and a (capito-) dorsoscapularis. Since the former muscle is inserted into the ventral extremity of the clavicle, into the sternum, and into the episternum (see Luther and Lubosch 1938), it is therefore comparable to the mammalian sternocleidomastoid. The (capito-) dorsoscapularis arises from the head caudally to the upper thoracic region and is inserted into the scapula. In some species the (capito-) dorsoscapularis tends to split into a cranial and a dorsal portion, while in others the cranial part may be entirely absent (Luther and Lubosch 1938).

The (capito-) dorsoscapularis of sauropsides seems to be the precursor of the trapezius muscular group in mammals. It is generally considered that the mammalian trapezius group comprises a capitoclavicular, a cervicoscapular, and a dorsoscapular portion which may either remain separate or fuse, thereby forming a common muscular plate. However, the fused variant of the trapezius group exists only in marsupials, prosimians, and primates (Luther and Lubosch 1938). On the basis of findings in insectivores which lack the capitoclavicular part of the trapezius, Streissler (1900) proposed that the mammalian trapezius group comprises only two muscles, namely, the dorsoscapularis superior and inferior. Since the capitoclavicular portion is the most variable muscle of the spinal accessory muscles in several mammals (Streissler 1900; Luther and Lubosch 1938; Jouffroy 1971), it was characterized by Streissler (1900) as an individual muscle termed "cleido-occipital." In this context, it is interesting to mention that the cleido-occipital (CLO) muscle and the cranial attachment of the superior trapezius (TRAP) are reciprocally related: if an occipital origin of the superior TRAP is absent, there exists a broad and distinct CLO, and vice versa (Streissler 1900; Jouffroy 1971). Therefore, the capitoclavicular bundles of the TRAP, if present in mammals, are thought to represent a CLO fused with the superior TRAP (Streissler 1900). In the rat, and likewise in many mammals, the CLO and the superior TRAP are not fused but separated by the omotransversarius muscle (Streissler 1900; Schück 1913; Luther and Lubosch 1938). In these animals the CLO appears to be part of the sternocleidomastoid group rather than of the trapezius group. Accordingly, Streissler (1900) proposed that the CLO belongs to the former group, its fusion with the TRAP depending merely on the involution of the omotransversarius during evolution. In consequence, the sternocleidomastoid group in mammals is regarded as consisting of three muscles, namely, the STM, CLM, and the variable CLO.

If one considers the phylogenetic development of the spinal accessory muscles, some information concerning the problem of the somatotopic organization of the spinal XI nucleus can be obtained. The first appearance of a trapezius in those fishes with a cartilaginous shoulder girdle suggests that the phylogenetically oldest muscle of the spinal accessory musculatur – the dorsoscapularis – is primarily concerned with movements of the scapula, i.e., elevation of the shoulder girdle. The progressive extension of the trapezius toward the head in amphibians and the insertion of this cranial portion into the acromial process of the scapula in claviculated anurans apparently indicate the derivation of the CLO from the trapezius muscular sheet. There exists, therefore, an intimate phylogenetic, topographic, and functional relationship between these two muscles acting on the shoulder girdle. Accordingly, the location of the motoneurons innervating both muscles – the TRAP and CLO – within the same subnucleus of the spinal XI nuclear complex is consistent with that relationship.

The fact that the sternocleidomastoid group does not make its appearance in vertebrates below the reptiles strongly suggests that the spinal accessory musculature has developed from a single dorsal branchiomeric unit rather than from both a dorsal (trapezius) and a ventral (sternocleidomastoid) component (Straus and Howell 1936). If one assumes a specific functional correspondence between motoneurons and target muscles as well as their mutual interrelationship during development, the derivation of the sternocleidomastoid from the trapezius muscles lends support to the belief that the neurons of the medial subnucleus have migrated from a common spinal XI nucleus into their medial position.

Coincidently, in sauropsides, certain marked changes appear first in the phylogenetic scale. The neck elongates, the shoulder girdle descends, and the secondary sternal elements appear concurrently with a sternocleidomastoid homologue and an increased ability to rotate the head. The concurrence of these developmental processes suggests that there may be a common need for their elaboration, most probably the need for an increased mobility of the head. Furthermore, the appearance of a sternocleidomastoid in amniotes depends obviously on the development of sternal elements rather than of the clavicle.

The changes taking place in the spinal accessory musculature of mammals seem to be only secondary phenomena leading merely to a regrouping of the preexisting muscles. Thus, in the rat, both the CLM and CLO muscles are attached to the middle third of the clavicle. However, compared to the attachment of the CLO, that of the CLM is slightly shifted toward the sternal extremity of the clavicle. The tendency of these muscular attachments to shift in opposite directions, namely, that of the CLM toward the sternum and that of the CLO toward the acromion, is present in several mammals (Streissler 1900). In primates, including man, the migration of these two muscles toward opposite extremities of the clavicle is quite definite, the CLM being united with the STM and the CLO with the TRAP. In view of the developmental shifts referred to above, it is justifiable to assume that in the mammalian scale the motoneurons of the CLM tend to migrate progressively into a medial position, thereby joining the neurons of the STM. In the rat, these migrations of the CLM muscle and of its motoneurons into a medial position seem to be at an orimentary stage. The predominant localization of the motoneurons innervating the CLM within the lateral subnucleus may reflect the primary action of the CLM on the shoulder

36

girdle synergically with the CLO and TRAP. However, the tendency of the CLM neurons to migrate toward the medial subnucleus seems to indicate a functional shift of the muscular action taking place in the rat. Since mainly neurons innervating the STM are localized in the medial subnucleus, the prospective function of the CLM muscle may be movement of the head and neck.

The migration of motoneurons from the rostral part of the spinal XI nucleus into a medial position therefore seems to depend on a functional change of some spinal accessory muscles in the course of evolution. The question, however, remains: why do these neurons migrate into a medial position? Conceivably, subgrouping of neurons increases the delicacy of their control, but this does not explain the displacement of the motoneurons in a particular direction. We submit that the direction of the neuronal displacement may be determined by particular changes in the sources of the neurons' most important stimulation, most probably by specific changes in their supraspinal control.

4.2.3.2 Supraspinal Control of the Medial Subnucleus of the Spinal XI Nuclear Complex

Let us consider the position of the medial subnucleus of the spinal XI nuclear complex in conjunction with the site of termination of descending tracts from supraspinal levels. The dorsomedial part of the ventral horn receives terminals from many supraspinal pathways, particularly from those which, classified as the medial descending system (Kuypers 1964), most directly influence trunk and proximal extremity musculature. Some of these descending pathways terminate, either predominantly or exclusively, in the cervical cord. These latter fiber systems descend within the sulcomarginal zone, i.e., in an area bounded laterally by the medial margin of the ventral horn, medially by the ventral median fissure, and dorsally by the ventral white commissure. The sulcomarginal zone, considered as the spinal prolongation of the medial longitudinal fasciculus, contains four fiber systems connecting some subcortical centers with the cervical cord in particular, namely, the tectospinal, interstitiospinal (Cajal), medial vestibulospinal, and pontine reticulospinal tracts (Busch 1961, 1964; Schoen 1964). There is anatomical evidence from experimental studies that several sulcomarginal tracts terminate in the dorsomedial part of the ventral horn of the cervical cord (Nyberg-Hansen 1964, 1965, 1966a, b; Nyberg-Hansen and Mascitti 1964; Petras 1967). This region of the ventral horn, corresponding to the commissural nucleus (J.v. Lenhossék 1855; M.v. Lenhossék 1889, 1895; Ramón y Cajal 1909) and to the centrodorsal nucleus of Rexed's lamina VIII, is thought to contain no motoneurons and to be purely propriospinal (Rexed 1952, 1954). However, this general assumption is no longer tenable, as we learned that motoneurons innervating neck muscles – epaxial muscles (Richmond et al. 1978; Robards et al. 1980), hypobranchial muscles (Gottschall et al. 1980a), and spinal XI muscles (Gottschall et al. 1980b; Robards et al. 1980; Karim and Hoo Nah 1981) – are located within these areas. In addition, evidence from stimulation experiments indicates that several sulcomarginal tracts are linked monosynaptically to the motoneurons innervating neck muscles (Wilson and Yoshida 1969; Anderson et al. 1971; Wilson and Maeda 1974; Fukushima et al. 1979; Peterson 1979).

The importance of these descending pathways terminating on neck motoneurons in the control of head and neck movements has long been recognized, particularly with regard to labyrinthine influences on postural reflexes and the coordination of head and eye movements during orienting and visual tracking (see Coulter et al. 1979). Moreover, stimulation of the pontine reticulospinal, tectospinal, and interstitiospinal (Cajal) tracts has been shown to produce direction-specific movements of eyes and head. Thus, in the horizontally postured cat, ipsiversive turning is induced by stimulation of the pontine reticulospinal formation (see Podivinský 1968; Peterson 1979), whereas contraversive turning is elicited by stimulation of the optic tectum (see Podivinský 1968; Anderson et al. 1971). Electrical stimulation of the interstitial nucleus (Cajal) causes rotatory gaze shifts and head movements (Hyde and Eason 1959; Fukushima 1979; see also Podivinský 1968). Furthermore, the interstitiospinal tract has been shown to have a direct excitatory action on the motoneurons innervating the sternocleidomastoid in cats (Fukushima 1979).

Coordinate eye-head movements are especially important in those horizontally postured quadrupedal mammals that possess a rather limited oculomotor range. For example, cats cannot fixate, without moving their heads, a visual target whose angular distance from the area centralis exceeds 25°. To orient toward targets lying outside this range, these animals must necessarily use coordinate eye-head movements. Moreover, most of their eye saccades are accompanied by head movements in the same direction, even if the target lies within the reach of the eye alone (Guitton et al. 1980; Roucoux et al. 1980). Thus, it appears that the superior colliculus is intimately involved in the elaboration of visually guided gaze shifts by means of eye-, head-, and probably body-orienting movements. Body-turning and rotatory movements would be necessary in cases where the visual target cannot be foveated by an eye-head movement. In this regard, it is of interest that the tectum of animals whose head mobility is quite limited, such as amphibians and some reptiles, directly controls body movements (Ewert 1967; Shapiro and Goodman 1969).

In consequence, we propose that motoneurons of muscles, involved in visually guided orienting movements, have to lie, either primarily or secondarily, in the medial part of the ventral horn in order to receive the appropriate stimulation by descending sulcomarginal tracts. The evidence from the functional significance of these supraspinal pathways terminating in the medial part of the ventral horn in the cervical cord suggests that they mediate mainly coordinate eye-head movements in response to various stimuli. Furthermore, the motoneurons lying in the medial part of the ventral horn in the cervical cord supply mostly head movers, such as the STM and splenius which in turn have been shown to be mainly or entirely under supraspinal control (see e.g., Rapoport 1979; Robards et al. 1980).

When the facts listed above (to which many more could be added, if we were to treat the supraspinal control of the neck muscles more comprehensively) are evaluated, it appears to be natural to find the motoneurons supplying the STM, a prime mover of the head, in a medially "displaced" part of the spinal XI nucleus. If this is accepted, the development of the sternocleidomastoid muscular group and of a medial subnucleus in amniotes may be interpreted as a specific adaptation of the spinal XI nerve and its musculature to a functional need for an increased and more independent, but coordinated, mobility of the head. Apparently, these functional specializations meet the demand of amniotes to coordinate their eye-head movements toward targets in response to various stimuli. Assuming that the sternocleidomastoid and its motoneuronal subgroup were established in amniotes in accordance with their need for goal-directed head movements, one is inclined to consider that the trapezius of anamnia is concerned with locomotion, i.e., body movements, toward targets in response to various stimuli. In conclusion, we propose that the organization of the vertebrate spinal XI nerve and its musculature is inseparably connected with the performance of goal-directed and coordinated movements of eye, head, and body.

Since the spinal XI nucleus is intimately connected to various efferent nuclei and under the influence of sensory inputs of cranial nerves, the contention that the spinal XI nerve is a cranial nerve seems corroborated. Other arguments in favor of its cranial origin, particularly the proposed special visceral derivation of the spinal XI nerve, are based on the peculiar intramedullary course and emergence of its rootlets. The following discussion will deal more fully with this problem.

4.3 The Spinal Accessory (XI) Nerve: A Cranial Nerve of Special Visceral Origin?

Although situated in the cervical cord, the spinal XI nucleus is commonly regarded as a cranial motor nucleus (Fürbringer 1897; Gegenbaur 1898; Lubosch 1899; Black 1917a, b, 1920, 1922; Kimmel 1940). Architectonically, it differs indeed from other motoneuronal arrangements in the cervical cord. Whereas motoneurons innervating somitic neck muscles are widely and diffusely distributed throughout the ventral horn (Richmond et al. 1978; Gottschall et al. 1980a), the spinal XI neurons are clustered within two discrete subnuclei. This tight organization of motoneurons is a characteristic feature of efferent cranial nuclei (Krammer et al. 1979) and does not exist in the cervical cord, the phrenic nucleus being the only exception. The compact texture of the phrenic nucleus (Webber et al. 1979; Kuzuhara and Chou 1980; Rikard-Bell and Bystrzycka 1980; Gottschall 1981) is believed to be related to the need for cell synchronization during the inspiratory phase of respiration, which is mediated by synchronous inputs from medullary respiratory neurons (see Webber et al. 1979; Rikard-Bell and Bystrzycka 1980). As segregated nuclei under supraspinal control, the phrenic and spinal XI nuclei differ from several motoneuronal pools in the cervical cord.

As the organization of the spinal XI nucleus indicates its peculiarity among the motoneuronal pools in the cervical cord, so the developmental history of the nucleus suggests its cranial origin. In reptiles (Lubosch 1899; Black 1920) as in anamnia (Black 1917a, b), the spinal component of the accessory nucleus is generally thought to be located in the caudal end of the dorsal vagus column. This assumption is based on the observation that in those reptiles which have limbs, there exists a definite prolongation of the dorsal vagus column into the cervical cord. In apodal serpentes (such as the boa) in which the spinal XI nerve is absent, this cervical extension is also lacking, and the dorsal vagus nucleus accordingly terminates abruptly in the lower medulla (Black 1920; see also Straus and Howell 1936). Therefore, the above cervical extension is regarded as a true spinal XI nucleus.

During phylogeny, as the shoulder girdle drops and neck and spinal XI muscles lengthen, the peripheral end and the perikarya of the spinal XI nerve extend lower in the neck and the cervical cord, respectively (see Pearson et al. 1964). The motoneurons of the spinal XI nerve separate from the cranial associated nucleus and migrate from the dorsal vagus nucleus in a caudoventral direction to become located in the ventral horn of the cervical cord. This phylogenetic migration (Ariëns Kappers 1914, 1927; Ariëns Kappers et al. 1936) of the spinal XI nucleus seems to be repeated in its ontogenetic development. Thus, ontogenetic studies have shown that the spinal XI nucleus in both birds (see Black 1922) and mammals (Kimmel 1940) originates from the dorsal vagus nucleus and migrates from this primitive position to its definitive location in the ventrolateral part of the ventral horn. The close association of the spinal XI nucleus and the nucleus ambiguus which occurs in some mammals was regarded by Ariëns Kappers (see Straus and Howell 1936) as a secondary development resulting from a similar ventrolateral descent of the nucleus ambiguus.

Coincident with the increased role of the spinal XI muscles as we ascend in the vertebrate scale, we note in amniotes a great advance in the development of the spinal XI nerve, which obtains additional rootlets of origin by spreading progressively caudalward (Streeter 1908). Among the "neurobiotactic" (Ariëns Kappers 1914, 1927; Ariëns Kappers et al. 1936) influences which cause the spinal XI nucleus to migrate into the cervical cord, one should consider the codescent and cofunctioning of the spinal XI muscles with other neck muscles and the stimulation of the respective motoneurons coming from similar sources (Black 1917a; Pearson et al. 1964).

According to Ariëns Kappers (1914, 1927) and Kappers et al. (1936), positive evidence of a displacement of a nucleus from its more primitive embryological and phylogenetic position into a definite location may be adduced when a motor root takes a more or less indirect emergent course, the direction of the nuclear displacement being indicated by the course of the emerging fibers. As regards the spinal XI axons, their emergent routes are bent with reference to vertical and transverse planes, thus suggesting a nuclear displacement in two dimensions. Concerning the spinal XI fasciculus, a caudal migration of the spinal XI nucleus appears to be indicated by the intramedullary ascent of the spinal XI axons. Moreover, being unique in the spinal cord, this ascending course of motor fibers is suggestive not only of a nuclear descent but also of a cranial origin of the spinal XI nucleus. This intramedullary ascent of the spinal XI axons, though it had been recognized by some authors (Turner 1895; Koelliker 1896; Hepburn and Waterston 1904; Pearson 1938; Crosby et al. 1962; Pearson et al. 1964), had been taken only by Pearson et al. (1964) to support the idea of a caudal displacement of the spinal XI nucleus. Lack of experimental evidence is one of the probable reasons for the fact that such a peculiar feature of the spinal XI nerve had been overlooked by most investigators. In addition, the spinal XI fasciculus had been mistaken for a descending spinal part of the solitary tract, known as the "respiratory bundle of Krause," as was pointed out by some authors (Turner 1895; Koelliker 1896; Hepburn and Waterston 1904). However, our present experiments clearly demonstrate that the fascicle, which ascends in the formatio reticularis, is formed by fibers arising in the spinal XI nucleus.

As the ascending course of the spinal XI axons suggests a caudal migration of an originally cranial nucleus, so its special visceral derivation seems to be indicated by the indirect emergent route of the spinal XI axons with respect to transverse planes (see e.g., Lubosch 1899). These fibers course dorsomedially toward the afferent area of the cervical cord before turning lateralward to emerge from the lateral surface of the spinal cord. This path taken by the spinal XI axons closely resembles the indirect emergent course of fibers arising in special visceral motor nuclei, particularly in the nucleus ambiguus (e.g., Kimmel 1940; Crosby et al. 1962). As stated by Ariëns Kappers (1914, 1927) and Kappers et al. (1936), the dorsomedial route of the axons that arise in the nucleus ambiguus reflects a ventrolateral displacement of this nucleus in the course of ontogenetic development. On the basis of the analogy of the emergent course of the spinal XI fibers, we assume a similar ventrolateral migration of the spinal XI nucleus. Since a ventrolateral displacement occurs characteristically in cranial nuclei of the special visceral category (see e.g., Black 1917a; Kimmel 1940; Crosby et al. 1962; Heaton and Moody 1980), the peculiar emergent course of the spinal XI fibers therefore supports the view that the spinal XI nerve is a branchiomeric nerve.

Further evidence in support of a cranial origin of the spinal XI nucleus is derived from our results concerning the significance of the communicating branches of the cervical plexus. In the rat as in various mammals (Straus and Howell 1936; Corbin and Harrison 1938; Yee et al. 1939), these branches of the cervical plexus have been shown to contribute solely proprioceptive fibers to the spinal XI muscles. Via extraspinal anastomoses these branches communicate with purely motor branches of the spinal XI nerve, thereby forming mixed nerves. These peripheral anastomoses between purely motor and sensory components are uncommon in muscular branches of spinal origin but closely resemble those occurring in cranial nerves of the somatic efferent category.

However, the spinal XI nerve appears to be originally a mixed nerve, as this additional supply of the spinal XI musculature by cervical branches seems to be limited to higher vertebrates, i.e., sauropsides and mammals (see e.g., Addens 1933; Straus and Howell 1936). Moreover, embryological evidence clearly indicates the presence of a sensory component in the spinal XI nerve, even in mammals (Windle 1931b; Pearson 1938). These afferent fibers have been shown to arise in small intra- and extracranial ganglia and in scattered cell groups, being present along the course of the spinal XI nerve in many mammalian embryos, including human ones (Weigner 1901; Streeter 1905; Fahmy 1927; Windle 1931b; Pearson 1938; Kimmel 1940; Waibel 1954; Pearson et al. 1964). In contrast to dorsal root fibers, these sensory fibers of the spinal XI nerve pass ventral to the substantia gelatinosa to enter the solitary tract and the posterior funiculus (Windle 1931a, b; Pearson 1938; Pearson et al. 1964). Although in adult mammals the number of ganglion cells is too small to account for a proprioceptive innervation of the spinal XI muscles (Windle 1931b; Corbin and Harrison 1938), the occurrence in young embryos of sensory fibers joining the solitary tract suggests the presence of a visceral sensory component in this nerve. Apparently, the spinal XI nerve tends to lose its sensory component in the course of development. Embryological investigations by Streeter (1905) indicated that many of the original ganglion cells of the spinal XI nerve become incorporated in the dorsal root ganglia of cervical nerves (see also Straus and Howell 1936).

Available data are thus strongly supportive of the thesis that the spinal XI nerve is an originally mixed nerve of special visceral origin whose sensory fibers shift to a spinal pathway during ontogeny and phylogeny.

Finally, some additional anatomical facts regarding the emergence of the spinal XI rootlets have to be mentioned. Like some previous investigators (Henle 1871; Kazzander 1891; Lubosch 1899), we find that the emergent routes of the spinal XI rootlets shift in a dorsal direction in rostral cervical segments. Thus, the farther the spinal XI fibers are traced rostralward in the cervical cord, the more they approach during their intramedullary course the dorsal horn and the substantia gelatinosa and, furthermore, the entrance zone of the dorsal spinal roots at their exits. Although this dorsal displacement of the spinal XI rootlets occurs merely relative to the enlargement of the dorsal horn and substantia gelatinosa at rostral cervical levels, as mentioned by Lubosch (1899), this intimate relationship of emerging spinal XI rootlets and entering dorsal roots has been subject to misinterpretation. In fact, these rostrally emerging spinal XI rootlets have been considered as efferent fibers which pass as so-called fibers of Lenhossék, conjointly with the sensory fibers of the dorsal roots, to an unknown peripheral distribution (M.v. Lenhossék 1890, 1895; Ramón y Cajal 1890a, b, 1893; Gehuchten 1893, 1900).

Another point relates to the mixing of nerve fibers in the intramedullary spinal XI fasciculus. This intermingling of spinal XI axons occurs in such a manner that the nerve fibers constituting a particular rootlet arise at various distances caudal to their exit. Although the exact distance of ascent of individual fibers is difficult to determine, it seems highly probable that they ascend for more than one cervical segment. Accordingly, the spinal XI rootlets appear to be of plurisegmental origin, a condition not usually found in motor roots of spinal nerves. However, a corresponding plurisegmental arrangement of spinal nerves is present in the limb plexuses, being formed by the union of adjoining ventral branches from spinal nerves of different segments. But there are, undoubtedly, important differences with respect to the manner in which the plurisegmental arrangements are formed. While the formation of limb plexuses occurs extraspinally, the spinal XI axons are intermingled intramedullary in the ascending fasciculus. Furthermore, the ventral rami of spinal nerves which are intermingled in limb plexuses, are, in general, mixed nerves. On the contrary, in the spinal XI nerve, the plurisegmental arrangement is formed by purely motor fibers, the other components being added by branches of cervical ventral rami via peripheral anastomoses. As compared with the formation of limb plexuses, the plurisegmental arrangement of motor fibers is shifted in the spinal XI nerve toward the central nervous system. In contrast, the formation of mixed muscular branches is displaced far toward the periphery, since it depends on the communication of the spinal XI nerve with spinal nerves. Therefore, in the spinal XI nerve, the formation of plurisegmental and mixed muscular branches seems to be split up into two spatially successive processes which occur simultaneously in the limb plexuses.

In conclusion, the organization of the spinal XI nerve differs in so many aspects from that of muscular nerves of spinal and somatic cranial origin that this nerve can hardly be classified as either.

4.4 Significance of the Motoneuronal Organization in the Cervical Cord

Our present understanding of the motoneuronal organization in the cervical cord is based mainly on cytoarchitectonic studies unsupported by experimental evidence. Views concerning the possible functional significance of the motoneuronal groups in the cervical cord are almost entirely extrapolated from experimental results observed chiefly in limb areas of the spinal cord. Thus, in limb enlargements of the cord, the ventromedial group of motoneurons is thought to innervate axial musculature (supplied by dorsal primary rami), whereas the limb muscles are considered to be supplied by ventral primary rami derived from the lateral neuronal groups (Kaiser 1891; Marinesco 1898; Angulo y González 1927, 1939; Goering 1928; Reed 1940; Romanes 1941, 1964; Elliott 1942, 1944; Baulac and Meininger 1979). In assuming a basically similar organization in all cord regions, it has been concluded that in general the ventromedial and lateral groups of neurons are sources of fibers entering the respective dorsal and ventral primary rami.

However, experimental results obtained by Sprague (1948, 1951) indicated that this classical concept of motor organization applies only to areas of limb plexuses, being invalid for the remainder of the cord. In fact, in nonlimb parts of the cord, the perikarya of both dorsal and ventral rami are widespread throughout most of the ventral horn. Although both neuronal groups overlap considerably, the dorsal ramus cells are more numerous in ventromedial parts and in the ventral tip of the ventral horn; in contrast, the perikarya of the ventral rami are more numerous in dorsolateral parts of the ventral horn (Sprague 1948). Furthermore, Sprague (1948) found the motoneurons of dorsal and ventral primary rami in a region not previously thought to contain motor cells, namely, at the medial border of the ventral horn near the ventral commissure (J.v. Lenhossék 1855; M.v. Lenhossék 1889, 1895; Ramón y Cajal 1909).

Despite these striking experimental findings, a subsequent concept of spinal cord organization (Rexed 1952, 1954, 1964), though deduced from merely structural data, is the most widely used. During the past few years, interesting results have been obtained through the use of the retrograde transport of HRP to identify particular motoneuronal groups within the cervical cord. These findings conflict seriously with Rexed's scheme of the motoneuronal organization in the cervical cord. Thus, the present study necessitates considerable revision of Rexed's assumption (1952, 1954) of the spinal XI nuclear location. In addition, it is obvious from HRP experiments that the perikarya of various neck muscles are present in those regions of the ventral horn which are thought to be propriospinal. Thus, in the dorsomedial region of the ventral horn, corresponding to the "commissural" and "centrodorsal" nuclei of Rexed's lamina VIII (Rexed 1952, 1954, 1964), are found the medial subnucleus of the spinal XI nuclear complex as well as motoneurons supplying dorsal neck and infrahyoid muscles (Rapoport 1978; Richmond et al. 1978; Gottschall et al. 1980a, b; Robards et al. 1980; Karim and Hoo Nah 1981). The perikarya supplying dorsal neck and infrahyoid muscles are additionally identified along the entire medial border of the ventral horn, in the poorly defined interspace between the ventromedial "nucleus" and the lateral subnucleus of the spinal XI nuclear complex, as well as throughout Rexed's propriospinal lamina VIII (Richmond et al. 1978; Gottschall et al. 1980a; Robards et al. 1980).

Inasmuch as these recent experimental results are at variance with Rexed's concept of motoneurons located exclusively within lamina IX (Rexed 1952, 1954, 1964), they corroborate Sprague's original assumption that the perikarya of both the ventral and the dorsal primary rami are widely distributed throughout the ventral horn, including the dorsomedial "commissural" region (Sprague 1948). Moreover, the tendency of the cells supplying infrahyoidal muscles to lie dorsal relative to those innervating dorsal neck muscles (Richmond et al. 1978; Gottschall et al. 1980a) confirms, in general, Sprague's view that the ventral ramus cells are localized dorsal to the dorsal ramus cells.

As these results can hardly be reconciled with Rexed's prevailing scheme of spinal motor organization, we propose a new concept concerning the possible significance of the motoneuronal organization in the ventral horn of the cervical cord.

Even though the motoneurons innervating somitic neck muscles are pooled and have a widespread distribution (Richmond et al. 1978; Gottschall et al. 1980a) as compared with the distinct and segregated spinal XI subnuclei, their position within the ventral horn, nevertheless, appears to be geometrically organized. Although such an organization seems to be inherent in the ventral horn, this special order of the neuronal groups is, because of their considerable overlap, not obvious without the use of experimental procedures.

Basically, the spinal motoneurons appear to be arranged relative to the surface as well as to radial axes of the ventral horn. Consequently, we suggest that two different patterns of spatial order exist in the ventral horn: a radial geometry, whereby the ventral horn may be subdivided into virtual consecutive sectors, and, secondly, stratification corresponding to which its motoneurons seem to be arranged in two concentric layers (Fig. 11). We use these two fundamental and superimposed patterns of neuronal arrangement as a frame of reference in respect of proposals to equate these patterns with somatotopic ideas.

Evaluation of the location and spatial relationship of motoneuronal groups reveals that the particular position of a neuronal group depends on certain features of the innervated muscle, e.g., the position of the muscle with respect to the skeleton. Moreover, those neuronal groups supplying muscles which share a particular feature appear to be localized within the same area with regard to the geometric reference grid. Therefore, we are inclined to assume that the pattern of motoneuronal organization is related to certain properties of the target muscles. Furthermore, the two geometric patterns of neuronal arrangement seem to correspond approximately to different features of the innervated muscles.

Thus, we believe that the stratification of the ventral horn into concentric laminae, namely, a marginal and an internal layer, reflects a somatotopic grouping of motoneurons innervating muscles which share the same position with respect to the axial and appendicular skeleton. This conclusion is based on the fact that neurons supplying hypaxial muscles lie in the medial part of the ventral horn more interiorly than those innervating epaxial muscles (Richmond et al. 1978; Gottschall et al. 1980a). It is corroborated by the observation that within the lateral cell group of the cervical intumescence, the perikarya innervating the posterior groups of limb muscles lie peripherally to those supplying

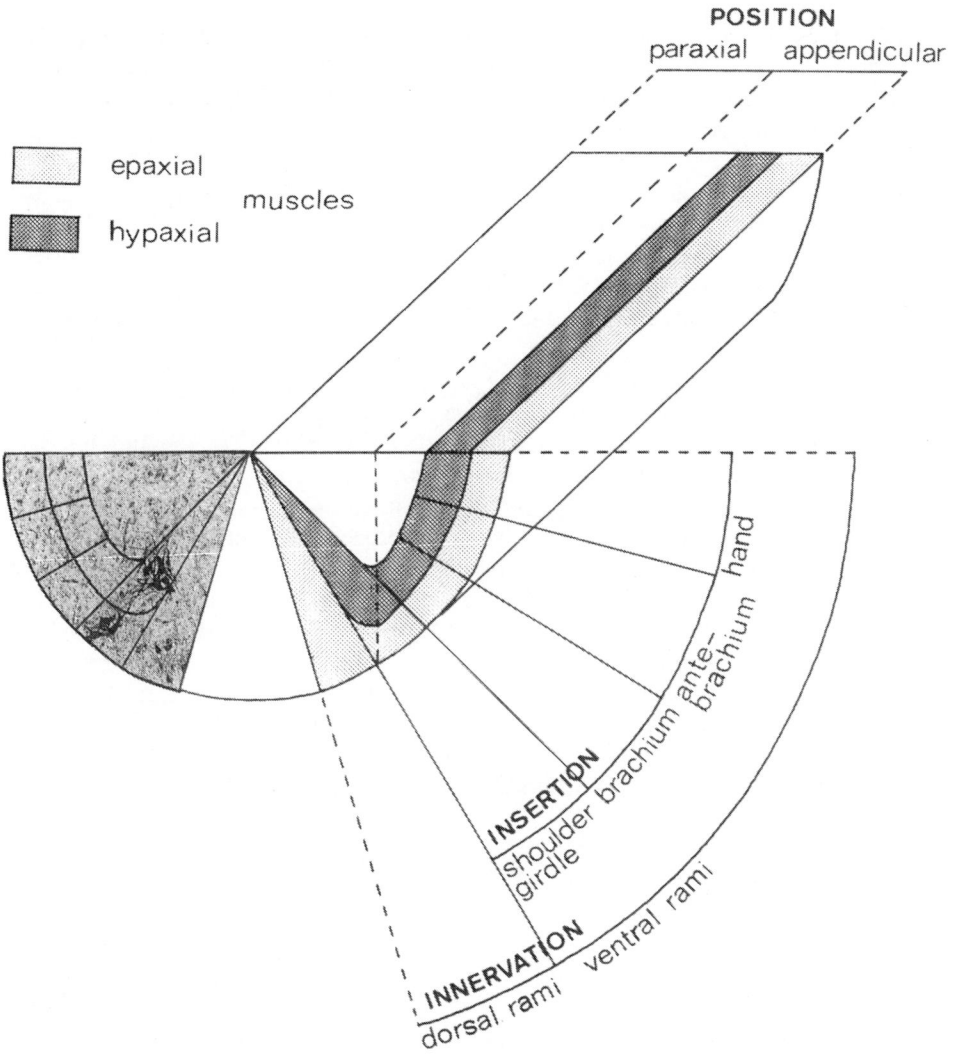

POSITION
paraxial   appendicular

epaxial
           muscles
hypaxial

INSERTION
shoulder girdle brachium ante-brachium hand

INNERVATION
dorsal rami ventral rami

Fig. 11. An abstract diagram of the ventral gray column (*right*), illustrating our concept of the significance of the motoneuronal organization in the cervical cord. Note the location of the labeled subnuclei of the spinal accessory (XI) nuclear complex, shown in the *inset* (*left*), within the geometric frame of reference (see Sect. 4.4 for detailed descriptions)

the anterior muscular groups (see e.g., Sprague 1948; Romanes 1964; Baulac and Meininger 1979). Since the anterior and posterior muscular groups of the upper limb function, respectively, as flexors and extensors, the hypothesis of a somatotopic grouping of the perikarya with respect to the function of the innervated muscles was advanced with some confidence (see e.g., Sprague 1948; Romanes 1964). This functional interpretation of the motoneuronal strata, though acceptable for motoneurons innervating limb muscles, does not however apply to the corresponding layers in the medial part of the ventral horn. There, the perikarya supplying hypaxial muscles, i.e., epibranchial and hypobranchial muscles, lie within the internal layer. As the hypaxial position of the infrahyoid

45

muscles does not concur with a flexion of the axial skeleton, it seems impossible to relate in general the stratification of the ventral horn to muscular function. Rather, we believe that the position of muscles, either dorsal or ventral with respect to the skeleton, correlates more comprehensively with the arrangement of motoneurons in the respective marginal and internal layers of the ventral horn. Therefore, we conclude that in the medial part of the ventral horn, the marginal and internal strata represent pools of neurons innervating the respective epaxial and hypaxial muscles. The neurons situated within the corresponding layers in the lateral part of the ventral horn are believed to innervate limb muscles lying in respective dorsal and ventral positions relative to the appendicular skeleton.

This laminar arrangement of the spinal motoneurons appears to be superimposed on another geometric pattern, i.e., a radial organization. These two patterns overlap in such a manner that they coincide largely in the medial part of the ventral horn, whereas they intersect in the lateral "appendicular" part, thereby being approximately perpendicular to each other.

Fictive radial axes are used as a system of reference to define the boundaries of adjacent sectors of the ventral horn within which neuronal pools of different significance are considered to be located. A radial line, drawn through the ventral horn from dorsomedial to ventrolateral, separates motoneurons which are sources of fibers entering dorsal primary rami from those entering ventral primary rami. The most medial sector includes the marginal zone of the commissural region, the entire medial margin, and also the ventral tip of the ventral horn. Within its confines lie perikarya giving rise to exclusively dorsal primary rami (see Sprague 1948; Richmond et al. 1978), whereas the motoneurons of several consecutive sectors are sources of ventral primary rami. Since the most medial sector of the ventral horn coincides with the medial part of the marginal stratum, the significance of the neuronal groups lying within this area can be concurrently accounted for in terms of dorsal primary rami and epaxial muscles.

In nonlimb regions of the cervical cord, the adjoining sector is the only dorsolateral zone of the ventral horn. It includes motoneurons which innervate hypaxial muscles by way of ventral rami (see Sprague 1948; Gottschall et al. 1980a). Hence, in nonlimb areas of the cord, there exists an additional coincidence of radial and laminar organization of the motoneurons, whereby the first dorsolateral sector corresponds to the internal layer of the ventral horn. With regard to the significance of the motoneuronal groups in the limb enlargement, this sector has to be divided by an arbitrary sagittal line into a larger medial and a smaller lateral part. As in nonlimb areas the perikarya lying in the medial part of this sector innervate hypaxial muscles. Within the lateral part, however, there occurs a somatotopic grouping of neurons innervating muscles which are inserted into the shoulder girdle. Since the radial and laminar organizations do not coincide in the lateral "appendicular" part of the ventral horn, the motoneuronal pools in the lateral part of this sector are stratified into a marginal layer, supplying dorsal "extensors", and an internal layer, innervating ventral "flexors" of the shoulder girdle.

The consecutive sectors, being maximally three in number, are present only in the limb enlargement of the cervical cord. We assume that the motoneuronal groups lying within these sectors represent somatotopic arrangements of neurons innervating muscles inserted into particular limb segments. Accordingly, the neurons supplying muscles inserted into the proximal part of the upper limb, i.e., into the humerus, are present in the most ventral of these three sectors. Furthermore, those perikarya innervating muscles inserted into the successively more distal segments of the limb are located within successively more dorsal sectors of the ventral horn. This preceding proposal is confirmed to some degree by experimental work (e.g., Romanes 1941, 1951, 1964; Baulac and Meininger 1979). Thus, cells innervating the more distal muscles in the limb have been found dorsal to those for proximal muscles. Because of the intersection of laminar and radial neuronal organization in those limb regions of the ventral horn, the perikarya of these sectors are further stratified into marginal cells, supplying posterior limb muscles, and internal cells, innervating anterior muscle groups.

In attempting to localize the subnuclei of the spinal XI nuclear complex within the geometric frame of reference, it is interesting to note that the medial subnucleus lies in the medial part of the ventral horn within its internal layer (Fig. 11). This position of these spinal XI motoneurons supplying mainly the STM corresponds very well to the innervation of hypaxial muscles by those neurons lying in this part of the internal stratum. However, the particular dorsomedial position of the medial XI subnucleus is shared only by perikarya innervating those infrahyoid muscles which are attached to the sternum (Gottschall et al. 1980a). Hence, a somatotopic subgrouping – with respect to muscular attachment – of the neurons supplying hypaxial muscles seems to occur in this part of the ventral horn. Thus, we suggest that hypaxial muscles with sternal insertion are supplied, irrespective of their embryological derivation, by motoneurons which lie in the dorsomedial part of the internal layer.

With regard to the radial reference system, the lateral subnucleus of the spinal XI nuclear complex lies in the same sector as its medial fellow, i.e., in the first dorsolateral sector (Fig. 11). However, in contrast to the medial subnucleus, its perikarya are situated in the lateral "appendicular" part and, furthermore, in the marginal layer of this sector. This position of the lateral XI subnucleus is consistent with our concept that muscles supplied by neurons located in this region are attached to the appendicular girdle and lie, at least in part, dorsal with respect to the skeleton. The existence of a cell grouping in this sector, with respect to the insertion of the innervated muscles, is corroborated by the distribution of the neurons supplying the omohyoid (Gottschall et al. 1980a). With regard to our frame of reference, both bellies of the omohyoid are innervated by motoneurons lying throughout the internal layer of this first dorsolateral sector. Both, the hypaxial position of these muscles and their innervation by ventral primary rami are in accord with this location of their perikarya. However, the neurons supplying the superior belly are found dorsomedially, and those of the inferior belly ventrolaterally. Furthermore, the motoneurons of the inferior omohyoid lie adjacent to the lateral XI subnucleus, but more interiorly. Therefore, not only the insertion of the inferior omohyoid into the shoulder girdle, but also the hypaxial and ventral appendicular position of this muscle are mirrored by this particular location of its perikarya. Interestingly, the neurons innervating the inferior omohyoid are replaced in more caudal cervical segments by the phrenic nucleus, which lies in a similar spatial relation to the lateral XI subnucleus. This location of the phrenic nucleus (Duron et al.

1979; Webber et al. 1979; Kuzuhara and Chou 1980; Rikard-Bell and Bystrzycka 1980), being analogous to that of neurons supplying the inferior omohyoid, further supports the view that the muscular diaphragm is derived from the hypobranchial muscular mass.

Finally, an attempt was made to relate several known facts to the conventional division of the ventral horn into medial and lateral parts, a division which is based entirely on the selection of some arbitrary sagittal line as a marker. The neurons found within the medial portion correspond to those of both layers of the ventral horn and therefore innervate several paraxial muscles, i.e., hypaxial and epaxial muscles. Furthermore, these perikarya give rise to both dorsal and ventral primary rami according to the presence of two sectors in this region. However, as laminar and radial organizations coincide within this medial part of the ventral horn, there is also a corresponding concurrence of the significance of the motoneuronal pools.

On the other hand, the motoneurons of the lateral part of the ventral horn are sources of exclusively ventral primary rami and supply entirely appendicular muscles. Since radial and laminar arrangements interesect in this "appendicular" part of the ventral horn, there is also a corresponding distinction of the significance of the neuronal groups. Thus, several sectors related to muscular attachments to particular limb segments are in addition arranged in neuronal strata supplying dorsal extensors and ventral flexors of the limb. Most probably, the marginal and internal strata of the "appendicular" portion of the cervical ventral horn are concurrently sources of the respective dorsal and ventral divisions of the brachial plexus. Thus, both portions of the ventral horn include neuronal pools of varying significance, whereas on the other hand, different spatial organizations exist in each of the portions. Therefore, this arbitrary subdivision of the ventral horn is not sufficient to distinguish neuronal groups of different significance.

Unescapably, the arrangement of the motoneurons within the ventral horn is neatly laid out according to a definite spatial plan. The question, however, remains: which agents operate in the formation and orientation of these neuronal groups? The whole matter yields some of its secrets as soon as one visualizes the motoneuronal pattern not merely descriptively, but rather from an ontogenetic viewpoint.

Thus, the neuronal pools of the ventral horn are placed in a particular order which correlates with the sequence of the concurrent muscular and motoneuronal differentiation during ontogenesis (Angulo y González 1927, 1939; Romanes 1941). When one considers the significance of the neuronal groups from medial to lateral and from ventrolateral to dorsolateral within each layer, there is no doubt that this spatial sequence of the motoneurons corresponds to the respective truncofugal and proximodistal succession of the differentiation of the innervated muscles (see e.g., Bardeen and Lewis 1901; Lewis 1902, 1910; Romanes 1941; Jouffroy 1971). The truncofugal sequence of muscular development, namely, the differentiation of the trunk muscles before the appendicular muscles, correlates therefore with the progressive development (Romanes 1941) and definite spatial sequence of the motoneurons from medial to lateral. Likewise, the proximodistal gradient of muscular differentiation in limbs (see e.g., Bardeen and Lewis 1901; Lewis 1902, 1910; Jouffroy 1971) coincides with the ventrodorsal succession of development in the lateral neuronal group (Romanes

48

1941). This developmental succession, in turn, is reflected by the ventrodorsal spatial sequence of the corresponding motoneuronal groups.

The stratification of the motoneurons into marginal and internal layers, on the other hand, seems to reflect the dorsoventral gradient of differentiation occurring in several muscular groups. Obviously, the motoneurons of the marginal layer innervate epaxial muscles and dorsal extensors of the limb. These muscles, lying dorsal with respect to the axial and appendicular skeleton, are known to differentiate before the muscles lying in corresponding ventral positions (see e.g., Bardeen and Lewis 1901; Lewis 1902, 1910; Jouffroy 1971). Accordingly, the neurons supplying hypaxial muscles and ventral flexors of the limb lie internal with regard to those innervating their dorsal fellows.

In conclusion, we suggest that the spatial order of the motoneurons in the ventral horn is the transcript of the chronological sequence of concurrent muscular and motoneuronal differentiation in ontogenesis.

Summary

Location and organization of the spinal accessory (XI) nucleus have been examined in the rat, using the technique of retrograde transport of horseradish peroxidase. The spinal XI nucleus comprises two distinct subnuclei, namely, a medial subnucleus present between the medullospinal transition and caudal levels of the 3rd cervical segment, and a lateral subnucleus extending from rostral levels of the 2nd to caudal levels of the 7th cervical segment. These subnuclei of the spinal XI nuclear complex, while overlapping in a longitudinal direction and occupying different positions within the ventral horn, fuse together at the caudal extremity of the medial subnucleus. The position of the spinal XI nuclear complex is constant throughout its rostrocaudal extent in the cervical cord, the apparent positional shift of the subnuclei being due merely to changes in the configuration of the ventral horn. The constancy of the nuclear position may be characterized by geometric principles. Thus, the long axis of a subnucleus is found at a constant distance from the central canal, measured along a virtual radial plane, which forms a constant angle with the median sagittal plane. This distance/angle relation, which characterizes the position of a nuclear long axis, is not only constant throughout the extent of a subnucleus but also peculiar to the subnucleus.

Nerve fibers arising in the caudal part of the lateral subnucleus form the origin of a spinal XI fasciculus which constantly receives axons from the spinal XI nuclear complex as it ascends intramedullary through the cervical cord. The caudalmost spinal XI rootlets make their exit from the spinal cord in the 5th cervical segment, thereby forming the origin of the extramedullary ascending spinal XI trunk. Spinal XI axons which emerge at the level of their perikarya follow an indirect double-bended route to exit from the spinal cord slightly ventral to those issuing at variable distances from the intramedullary ascending spinal XI fasciculus.

The spinal XI muscles receive their entire motor supply from axons arising in the spinal XI nuclear complex. No evidence for a motor component is found in the communicating branches of the cervical plexus by means of labeling or stimulation experiments. These branches of the cervical ventral rami, contributing proprioceptive fibers to the spinal XI muscles, arise in spinal ganglion cells of the 3rd–5th cervical root ganglia.

Each of the spinal XI subnuclei represents a conjoined motor pool for the innervation of at least two muscles. Thus, the sternomastoid and parts of the cleidomastoid are supplied by motoneurons of the medial subnucleus, whereas the neurons innervating the trapezius, cleido-occipital, and most of the cleidomastoid are pooled within the lateral subnucleus. Generally, an individual muscle is innervated by motoneurons confined to only one of the subnuclei, with the exception of the cleidomastoid whose neurons are located in those parts of both subnuclei which, facing one another, fuse together at the caudal extremity of the medial subnucleus. The significance of the somatotopic arrangement within the spinal XI nuclear complex has been evaluated on the basis of function and development of the innervated muscles, as well as from the viewpoint of the supraspinal control of the spinal XI motoneurons. A critical examination of the spatial and functional organization of the spinal XI nuclear complex has led us to a new concept concerning the significance of the motoneuronal organization in the cervical cord.

References

Abrahams VC (1977) The physiology of neck muscles; their role in head movement and maintenance of posture. Can J Physiol Pharmacol 55:332–338

Abrahams VC, Rancier F (1973) ATPase distribution in dorsal neck muscles of the cat. Can J Physiol Pharmacol 51:549–552

Addens JL (1933) The motor nuclei and roots of the cranial and first spinal nerves of vertebrates: I. Introduction. Cyclostomes. Z Anat Entwickl-Gesch 101:307–410

Allis EP, Jr (1897) The cranial muscles and cranial and first spinal nerves in Amia calva. J Morphol 12:487–808

Anderson ME, Yoshida M, Wilson VJ (1971) Influence of superior colliculus on cat neck motoneurons. J Neurophysiol 34:898–907

Angulo y González AW (1927) The motor nuclei in the cervical cord of the albino rat at birth. J Comp Neurol 43:115–142

Angulo y González AW (1939) The differentiation of the motor cell columns in the cervical cord of albino rat fetuses. J Comp Neurol 73:469–489

Ariëns Kappers CU (1914) Phenomena of neurobiotaxis in the central nervous system. Trans XVII Int Congr Med, London, 1913, Section I, Part II:109–122

Ariëns Kappers CU (1927) Three lectures on neurobiotaxis and other subjects. Acta Psychiatr Scand 2:118–185

Ariëns Kappers CU, Huber GC, Crosby EC (1936) The comparative anatomy of the nervous system of vertebrates, including man. Macmillan, New York

Bardeen CR, Lewis WH (1901) The development of the limbs, body wall and back. Am J Anat 1:1–35

Baulac M, Meininger V (1979) Organisation intramédullaire de la corne antérieure en fonction des nerfs du plexus brachial chez le rat. Étude par la peroxidase du raifort. Rev Neurol (Paris) 135:789–802

Beccari N (1913) Sulla spettanza delle fibre del Lenhossék al sistema del nervo accessorio e contributo alla morfologia di questo nervo. Arch Ital Anat Embriol 11:299–351

Beccari N (1914) Il IX, X, XI e XII pajo di nervi cranici e i nervi cervicali negli embrioni di Lacerta muralis. Arch Ital Anat Embriol 13:1–78

Black D (1917a) The motor nuclei of the cerebral nerves in phylogeny: a study of the phenomena of neurobiotaxis: I. Cyclostomi and Pisces. J Comp Neurol 27:467–564

Black D (1917b) The motor nuclei of the cerebral nerves in phylogeny: a study of the phenomena of neurobiotaxis: II. Amphibia. J Comp Neurol 28:379–427

Black D (1920) The motor nuclei of the cerebral nerves in phylogeny: a study of the phenomena of neurobiotaxis: III. Reptilia. J Comp Neurol 32:61–99

Black D (1922) The motor nuclei of the cerebral nerves in phylogeny: a study of the phenomena of neurobiotaxis: IV. Aves. J Comp Neurol 34:233–275

Bruce A (1901) A topographical atlas of the spinal cord. Williams and Norgate, London

Busch HFM (1961) An anatomical analysis of the white matter in the brain stem of the cat. Thesis, Royal Van Gorcum, Assen

Busch HFM (1964) Anatomical aspects of the anterior and lateral funiculi at the spinobulbar junction. In: Eccles JC, Schadé JP (eds) Organization of the spinal cord. Elsevier, New York, pp 223–237 (Prog Brain Res, vol 11)

Corbin KB, Harrison F (1938) Proprioceptive components of cranial nerves. The spinal accessory nerve. J Comp Neurol 69:315–328

Crosby EC, Humphrey T, Lauer EW (1962) Correlative anatomy of the nervous system. Macmillan, New York

Coulter JD, Bowker RM, Wise SP, Murray EA, Castiglioni AJ, Westlund KN (1979) Cortical, tectal and medullary descending pathways to the cervical spinal cord. In: Granit R, Pompeiano O (eds) Reflex control of posture and movement. Elsevier, New York, pp 263–279 (Prog Brain Res, vol 50)

Darkschewitsch L (1885) Über den Ursprung und den centralen Verlauf des Nervus accessorius Willisii. Arch Anat Physiol, Anat Abt, pp 361–378

Dawson AB (1925) The ossicle at the sternal end of the clavicle in the albino rat; the homologue of the sternal epiphysis of the clavicle in man? Anat Rec 30:205–210

DuBrul EL (1950) Posture, locomotion and the skull in lagomorpha. Am J Anat 87:277–315

Duron B, Marlot D, Larnicol N, Jung-Caillol MC, Macron JM (1979) Somatotopy in the phrenic motor nucleus of the cat as revealed by retrograde transport of horseradish peroxidase. Neurosci Lett 14:159–163

Edgeworth FH (1911) On the morphology of the cranial muscles in some vertebrates. Q J Microsc Sci 56:167–316

Edgeworth FH (1926) On the development of the coraco-branchiales and cucullaris in *Scyllium canicula*. J Anat 60:298–308

Ellenberger W, Baum H (1932) Handbuch der vergleichenden Anatomie der Haustiere. Springer, Berlin

Elliott HC (1942) Studies on the motor cells of the spinal cord: I. Distribution in the normal human cord. Am J Anat 70:95–117

Elliott HC (1944) Studies on the motor cells of the spinal cord: IV. Distribution in experimental animals. J Comp Neurol 81:97–103

Ewert JP (1967) Elektrische Reizung des retinalen Projektionsfeldes im Mittelhirn der Erdkröte (Bufo Bufo L). Pflügers Arch 295:90–98

Fahmy N (1927) A note on the intracranial and extracranial parts of the IXth, Xth and XIth nerves. J Anat 61:298–301

Favaro G (1903) Ricerche intorno allo sviluppo dei muscoli dorsali, laterali e prevertebrali negli amnioti. Arch Ital Anat Embriol 2:518–577

Frick H, Heckmann U (1955) Ein Beitrag zur Morphogenese des Kaninchenschädels. Acta Anat (Basel) 24:268–314

Froriep A (1882) Über ein Ganglion des Hypoglossus und Wirbelanlagen in der Occipitalregion. Arch Anat Physiol, Anat Abt, pp 279–302

Fukushima K, Hirai N, Rapoport S (1979) Direct excitation of neck flexor motoneurons by the interstitiospinal tract. Brain Res 160:358–362

Fürbringer M (1897) Über die spino-occipitalen Nerven der Selachier und Holocephalen und ihre vergleichende Morphologie. Festschr Gegenbaur 3:349–788

Gaupp E (1915) Das Schläfenbein und seine Darstellung im anatomischen, besonders im osteologischen Unterricht. Arch Anat Physiol, Anat Abt, pp 62–105

Gegenbaur C (1898) Vergleichende Anatomie der Wirbeltiere mit Berücksichtigung der Wirbellosen. Vol 1, Engelmann, Leipzig

Gehuchten A van (1893) Les éléments nerveux moteurs des racines posterieures. Anat Anz 8:215–223

Gehuchten A van (1900) Anatomie du systeme nerveux de l'homme. Vol 1, 3rd edn, Trois Rois, Louvain

Goering JH (1928) An experimental analysis of the motor-cell columns in the cervical enlargement of the spinal cord in the albino rat. J Comp Neurol 46:125–153

Gottschall J (1981) The diaphragm of the rat and its innervation. Muscle fiber composition; perikarya and axons of efferent and afferent neurons. Anat Embryol (Berl) 161:405–417

Gottschall J, Neuhuber W, Müntener M, Mysicka A (1980a) The ansa cervicalis and the infrahyoid muscles of the rat: II. Motor and sensory neurons. Anat Embryol (Berl) 159:59–69

Gottschall J, Zenker W, Neuhuber W, Mysicka A, Müntener M (1980b) The sternomastoid muscle of the rat and its innervation. Muscle fiber composition; perikarya and axons of efferent and afferent neurons. Anat Embryol (Berl) 160:285–300

Greene EC (1968) Anatomy of the rat. Hafner, New York

Guitton D, Crommelinck M, Roucoux A (1980) Stimulation of the superior colliculus in the alert cat: I. Eye movements and neck EMG activity evoked when the head is restrained. Exp Brain Res 39:63–73

Gura EV, Limanskii YuP (1977) Antidromic and synaptic potentials of motoneurons of the cat accessory nerve nucleus. Neurophysiology 8:246–248

Heaton MB, Moody SA (1980) Early development and migration of the trigeminal motor nucleus in the chick embryo. J Comp Neurol 189:61–99

Hebel R, Stromberg MW (1976) Anatomy of the laboratory rat. Williams and Wilkins, Baltimore

Henle J (1871) Nervenlehre. Vieweg, Braunschweig, Handbuch der systematischen Anatomie des Menschen. Vol 3(2)

Hepburn D, Waterston D (1904) A comparative study of the grey and white matter, of the motor-cell groups, and of the spinal accessory nerve, in the spinal cord of the porpoise (*Phocoena communis*). J Anat Physiol 38:105–118, 295–311

Hinsey JC, Corbin KB (1934) Observations on the peripheral course of the sensory fibers in the first four cervical nerves of the cat. J Comp Neurol 60:37–44

Hogg ID (1928) The motor nuclei of the cranial nerves of *Mus norvegicus albinus* at birth. J Comp Neurol 44:449–495

Holománová A, Beňuška J, Ďurkovičová C, Čierny G, Zlatoš J (1973) Localization of the motor cells after denervation of the sternocleidomastoid muscle in the cat. Folia Morphol (Praha) 21:335–337

Holománová A, Čierny G, Zlatoš J (1972) Localization of the motor cells of the spinal root of the accessory nerve in the cat. Folia Morphol (Praha) 20:232–234

Hömig J-P (1970) Die Anatomie des aktiven Bewegungsapparates der Albinoratte (*Mus rattus norvegicus albinos*). Dissertation, München

Hyde JE, Eason RG (1959) Characteristics of ocular movements evoked by stimulation of brainstem of cat. J Neurophysiol 22:666–678

Jouffroy F-K (1968) Musculature épisomatique. In: Grassé P-P (ed) Traité de Zoologie. Anatomie, systématique, biologie. Vol XVI (II). Masson, Paris, pp 479–548

Jouffroy F-K (1971) Musculature des membres. In: Grassé P-P (ed) Traité de Zoologie. Anatomie, systématique, biologie. Vol XVI (III). Masson, Paris, pp 1–476

Kaiser O (1891) Die Funktionen der Ganglienzellen des Halsmarkes. Nijhoff, Haag

Kampen PN van (1905) Die Tympanalgegend des Säugetierschädels. Morphol Jb 34:321–722

Karim MA, Hoo Nah Seang (1981) Localization of motoneurons innervating the sternocleidomastoid muscle in the monkey, rat and rabbit: a horseradish peroxidase study. Brain Res 206:145–148

Kazzander J (1891) Über den Nervus accessorius Willisii und seine Beziehungen zu den oberen Cervicalnerven beim Menschen und einigen Haussäugethieren. Arch Anat Entwickl-Gesch, pp 212–243

Kimmel DL (1940) Differentiation of the bulbar motor nuclei and the coincident development of associated root fibers in the rabbit. J Comp Neurol 72:83–149

Klima M (1968) Early development of the human sternum and the problem of homologization of the so-called suprasternal structures. Acta Anat (Basel) 69:473–484

Klima M (1973) The morphogenesis of the shoulder girdle and sternum in the monotremes (Mammalia: Prototheria). Adv Anat Embryol Cell Biol 47:6–80

Koelle GB, Friedenwald JS (1949) A histochemical method for localizing cholinesterase activity. Proc Soc Exp Biol Med 70:617–622

Koelliker A (1896) Handbuch der Gewebelehre des Menschen. Vol 2. Engelmann, Leipzig

Krammer EB, Rath T, Lischka MF (1979) Somatotopic organization of the hypoglossal nucleus: a HRP study in the rat. Brain Res 170:533–537

Kuypers HGJM (1964) The descending pathways to the spinal cord, their anatomy and function. In: Eccles JC, Schade JP (eds) Organization of the spinal cord. Elsevier, New York, pp 178–202 (Prog Brain Res, vol 11)

Kuzuhara S, Chou SM (1980) Localization of the phrenic nucleus in the rat: a HRP study. Neurosci Lett 16:119–124

Lenhossék J von (1855) Neue Untersuchungen über den feineren Bau des centralen Nervensystems des Menschen: I. Medulla spinalis und deren Bulbus rhachiticus. Denkschr math-naturwiss Classe Akad Wiss Wien 10:1–69

Lenhossék M von (1889) Untersuchungen über die Entwicklung der Markscheiden und den Faserverlauf im Rückenmark der Maus. Arch Mikr Anat 33:71–124

Lenhossék M von (1890) Über Nervenfasern in den hinteren Wurzeln, welche aus dem Vorderhorn entspringen. Anat. Anz 5:360–362

Lenhossék M von (1895) Der feinere Bau des Nervensystems im Lichte neuester Forschungen. Kornfeld, Berlin

Lewis WH (1902) The development of the arm in man. Am J Anat 1:145–183

Lewis WH (1910) Die Entwicklung des Muskelsystems. In: Keibel F, Mall FP (eds) Handbuch der Entwicklungsgeschichte des Menschen. Vol 1. Hirzel, Leipzig, pp 457–526

Lubosch W (1899) Vergleichend-anatomische Untersuchungen über den Ursprung und die Phylogenese des N. accessorius Willisii. Arch mikr Anat Entwickl-Gesch 54:514–602

Luther A, Lubosch W (1938) Muskeln des Kopfes: Viscerale Muskulatur. In: Bolk L, Göppert E, Kallius E, Lubosch W (eds) Handbuch der vergleichenden Anatomie der Wirbeltiere. Vol 5. Urban und Schwarzenberg, Berlin, pp 467–542, 1011–1106

Marinesco G (1898) Contribution à l'étude des localisations des noyaux moteurs dans la moëlle épinière. Rev Neurol (Paris) 6:463–470

Matsushita M (1970) The axonal pathways of the spinal neurons in the cat. J Comp Neurol 138:391–418

Mesulam M-M (1976) The blue reaction product in the horseradish peroxidase neurochemistry: incubation parameters and visibility. J Histochem Cytochem 24:1273–1280

Mesulam M-M (1978) Tetramethyl benzidine for horseradish peroxidase neurochemistry: a non-carcinogenic blue reaction-product with superior sensitivity for visualizing neural afferents and efferents. J Histochem Cytochem 26:106–117

Nyberg-Hansen R (1964) The location and termination of tectospinal fibers in the cat. Exp Neurol 9:212–227

Nyberg-Hansen R (1965) Sites and mode of termination of reticulo-spinal fibers in the cat. An experimental study with silver impregnation methods. J Comp Neurol 124:71–100

Nyberg-Hansen R (1966a) Functional organization of descending supraspinal fibre systems to the spinal cord. Anatomical observations and physiological correlations. Adv Anat Embryol Cell Biol 39:1–48

Nyberg-Hansen R (1966b) Sites of termination of interstitiospinal fibers in the cat. An experimental study with silver impregnation methods. Arch Ital Biol 104:98–111

Nyberg-Hansen R, Mascitti TA (1964) Sites and mode of termination of fibers of the vestibulospinal tract in the cat. An experimental study with silver impregnation methods. J Comp Neurol 122:369–389

Owen R (1868) On the anatomy of vertebrates. Vol III. Mammals. Longmans, London

Pearson AA (1938) The spinal accessory nerve in human embryos. J Comp Neurol 68:243–266

Pearson AA, Sauter RW, Herrin GR (1964) The accessory nerve and its relation to the upper spinal nerves. Am J Anat 114:371–391

Peterson BW (1979) Reticulospinal projections to the spinal motor nuclei. Annu Rev Physiol 41:127–140

Petras JM (1967) Cortical, tectal and tegmental fiber connections in the spinal cord of the cat. Brain Res 6:275–324

Podivinský F (1968) Torticollis. In: Vinken PJ, Bruyn GW (eds) Diseases of the basal ganglia. North-Holland, Amsterdam, pp 567–603 (Handbook of Clinical Neurology, vol 6)

Ramón y Cajal S (1890a) Sur l'origine et les ramifications des fibres nerveuses de la moëlle embryonnaire. Anat Anz 5:85–95, 111–119

Ramón y Cajal S (1890b) A quelle epoque apparaissent les expansions des céllules nerveuses de la moëlle épinière du poulet? Anat Anz 5:609–613, 631–639

Ramón y Cajal S (1893) Neue Darstellung vom histologischen Bau des Centralnervensystems. Arch Anat Physiol, Anat Abt, pp 319–428

Ramón y Cajal S (1909) Histologie du système nerveux de l'homme et des vertébrés. Vol I. Maloine, Paris

Rapoport S (1978) Location of sternocleidomastoid and trapezius motoneurons in the cat. Brain Res 156:339–344

Rapoport S (1979) Reflex connexions of the muscles involved in head movements in the cat. J Physiol (Lond) 289:311–327

Reed AF (1940) The nuclear masses in the cervical spinal cord of *Macaca mulatta*. J Comp Neurol 71:187–206

Rexed B (1952) The cytoarchitectonic organization of the spinal cord in the cat. J Comp Neurol 96:415–497

Rexed B (1954) A cytoarchitectonic atlas of the spinal cord in the cat. J Comp Neurol 100:297–379

Rexed B (1964) Some aspects of the cytoarchitectonics and synaptology of the spinal cord. In: Eccles JC, Schadé JP (eds) Organization of the spinal cord. Elsevier, New York, pp 58–92 (Prog Brain Res, vol 11)

Richmond FJR, Abrahams VC (1975) Morphology and enzyme histochemistry of dorsal muscles of the cat neck. J Neurophysiol 38:1312–1321

Richmond FJR, Scott DA, Abrahams VC (1978) Distribution of motoneurons to the neck muscles, biventer cervicis, splenius and complexus in the cat. J Comp Neurol 181:451–464

Rikard-Bell GC, Bystrzycka EK (1980) Localization of phrenic motor nucleus in the cat and rabbit studied with horseradish peroxidase. Brain Res 194:479–483

Robards MJ, Stritzel M, Robertson RT (1980) Ventral horn cells of the cervical cord project to neck muscles and brain. Brain Res. 189:519–523

Romanes GJ (1941) The development and significance of the cell columns in the ventral horn of the cervical and upper thoracic spinal cord of the rabbit. J Anat 76:112–131

Romanes GJ (1951) The motor cell columns of the lumbosacral spinal cord of the cat. J Comp Neurol 94:313–364

Romanes GJ (1964) The motor pools of the spinal cord. In: Eccles JC, Schadé JP (eds) Organization of the spinal cord. Elsevier, New York, pp 93–119 (Prog Brain Res, vol 11)

Roucoux A, Guitton D, Crommelinck M (1980) Stimulation of the superior colliculus in the alert cat: II. Eye and head movements evoked when the head is unrestrained. Exp Brain Res 39:75–85

Sandstrom CJ, Saltzman A (1944) A comparative study of the clavicular ligaments of the rat, rabbit, cat and dog. Anat Rec 89:23–32

Schoen JHR (1964) Comparative aspects of the descending fibre systems in the spinal cord. In: Eccles JC, Schadé JP (eds) Organization of the spinal cord. Elsevier, New York, pp 203–222 (Prog Brain Res, vol 11)

Schück A (1913) Beiträge zur Myologie der Primaten: II. 1. Die Gruppe: sterno-cleido-mastoideus, trapezius, omo-cervicalis. 2. Die Gruppe: levator scapulae, rhomboides, serratus anticus. Morphol Jb 47:355–419

Shapiro H, Goodman DC (1969) Motor functions and their anatomical basis in the forebrain and tectum of the alligator. Exp Neurol 24:187–195

Silver A, Wolstencroft JW (1971) The distribution of cholinesterase in relation to the structure of the spinal cord in the cat. Brain Res. 34:205–227

Sprague JM (1948) A study of motor cell localization in the spinal cord of the rhesus monkey. Am J Anat 82:1–27

Sprague JM (1951) Motor and propriospinal cells in the thoracic and lumbar ventral horn of the rhesus monkey. J Comp Neurol 95:103–123

Starck D (1967) Le crâne des mammifères. In: Grassé P-P (ed) Traité de zoologie. Anatomie, systématique, biologie. Vol XVI (I). Masson, Paris, pp 405–549

Straus WL Jr, Howell AB (1936) The spinal accessory nerve and its musculature. Q Rev Biol 11:387–405

Streeter GL (1905) The development of the cranial and spinal nerves in the occipital region of the human embryo. Am J Anat 4:83–117

Streeter GL (1908) The peripheral nervous system in the human embryo at the end of the first month. Am J Anat 8:285–303

Streissler E (1900) Zur vergleichenden Anatomie des M. cucullaris und M. sternocleidomastoideus. Arch Anat Entwickl-Gesch, pp 335–365

Turner WA (1895) The central connections and relations of the trigeminal, vago-glossopharyngeal, vago-accessory, and hypoglossal nerves. J Anat Physiol 29:1–15

Waibel P von (1954) Über das Vorkommen von Ganglienzellen in der Pars spinalis nervi accessorii. Acta Anat (Basel) 20:128–154

Webber ChL Jr, Wurster RD, Chung JM (1979) Cat phrenic nucleus architecture as revealed by horseradish peroxidase mapping. Exp Brain Res 35:395–406

Weigner K (1901) Beziehungen des Nervus accessorius zu den proximalen Spinalnerven. Anat Hefte 17:549–587

Weiss P (1969) Principles of development. Hafner, New York

Willis Th (1664) Cerebri Anatome, cui accessit Nervorum Descriptio et Usus. Londini: Typis Th Roycroft

Wilson VJ, Maeda M (1974) Connections between semicircular canals and neck motoneurons in the cat. J Neurophysiol 37:346–357

Wilson VJ, Yoshida M (1969) Monosynaptic inhibition of neck motoneurons by the medial vestibular nucleus. Exp Brain Res 9:365–380

Windle WF (1931a) The neurofibrillar structure of the spinal cord of the cat embryos correlated with the appearance of early somatic movements. J Comp Neurol 53:71–114

Windle WF (1931b) The sensory components of the spinal accessory nerve. J Comp Neurol 53:115–127

Yee J, Harrison F, Corbin KB (1939) The sensory innervation of the spinal accessory and tongue musculature in the rabbit. J Comp Neurol 70:305–314

Yoda S (1940) Über die Kerne der Medulla oblongata der Katze. Z Mikrosk Anat Forsch 48:529–582

Youssef EH (1966) The chondrocranium of the albino rat. Acta Anat (Basel) 64:586–617

Subject Index

61

Advances in Anatomy, Embryology and Cell Biology

Editors: F. Beck, W. Hild,
W. Kriz, R. Ortmann,
J. E. Pauly, T. H. Schiebler

Springer-Verlag
Berlin Heidelberg New York
London Paris Tokyo

Volume 101: A. Kress, J. Millian
The Female Genital Tract of the Shrew Crocidura russula
1987. 31 figures. VI, 76 pages. ISBN 3-540-16942-3

Volume 100: J. Altman, S. A. Bayer
The Development of the Rat Hypothalamus
1986. 125 figures. XV, 178 pages. ISBN 3-540-16654-8

Volume 99: W. F. Neiss
Ultracytochemistry of Intracellular Membrane Glycoconjugates
1986. 113 figures. VIII, 92 pages. ISBN 3-540-16726-9

Volume 98: K. Hinrichsen
The Early Development of Morphology and Patterns of the Face in the Human Embryo
1985. 82 figures. VII, 79 pages. ISBN 3-540-15848-0

Volume 97: D. K. Morest, J. A. Winer
The Comparative Anatomy of Neurons: Homologous Neurons in the Medial Geniculate Body of the Opossum and the Cat
1986. 43 figures. XI, 96 pages. ISBN 3-540-15726-3

Volume 96: C. L. Veenman, K.-M. Gottschaldt
The Nucleus Basalis-Neostriatum Complex in the Goose (Anser anser L.)
1986. 41 figures. VIII, 85 pages. ISBN 3-540-15338-1

Volume 95: M. Köhncke
The Chondrocranium of Cryptoprocta ferox
1985. 21 figures. VI, 89 pages. ISBN 3-540-15337-3

Volume 94: B. Demes
Biomechanics of the Primate Skull Base
1985. 29 figures. V, 59 pages. ISBN 3-540-15290-3

Springer

Advances in Anatomy, Embryology and Cell Biology

Editors: F. Beck, W. Hild, W. Kriz, R. Ortmann, J. E. Pauly, T. H. Schiebler

Springer-Verlag
Berlin Heidelberg New York
London Paris Tokyo